A Smoother Pebble

A Smoother Pebble

Mathematical Explorations

DONALD C. BENSON

UNIVERSITY PRESS

2003

OXFORD

UNIVERSITY PRESS

Oxford   New York
Auckland   Bangkok   Buenos Aires   Cape Town   Chennai
Dar es Salaam   Delhi   Hong Kong   Istanbul   Karachi   Kolkata
Kuala Lumpur   Madrid   Melbourne   Mexico City   Mumbai   Nairobi
São Paulo   Shanghai   Taipei   Tokyo   Toronto

Copyright © 2003 by Oxford University Press, Inc.

Published by Oxford University Press, Inc.
198 Madison Avenue, New York, New York 10016

www.oup.com

Oxford is a registered trademark of Oxford University Press

Library of Congress Cataloging-in-Publication Data
Benson, Donald C.
A smoother pebble : mathematical explorations / Donald C. Benson.
   p. cm.
Includes bibliographical references and index.
ISBN 0-19-514436-8
1. Mathematics—History—Popular works. I. Title.
QA21 .B46 2003
510—dc21   2002042515

9 8 7 6 5 4 3 2 1

Printed in the United States of America
on acid-free paper

Acknowledgments

First, I would like to thank my wife, Dorothy, for reading the entire manuscript many times and making many valuable suggestions.

Thanks to Ned Black and Donald Chakerian for reading parts of the manuscript and making valuable comments.

I would also like to acknowledge the assistance and encouragement of Kirk Jensen at Oxford University Press.

This book was typeset by the author using LaTeX. Many thanks to the authors of the LaTeX packages used here and all who have contributed to *CTAN*, the *Comprehensive TeX Archive Network*.

The line drawings were done by the author using MetaPost and Gnu-Plot.

Contents

II   The Shape of Things

III   The Great Art

IV   A Smoother Pebble

A Smoother Pebble

Introduction

I do not know what I may appear to the world; but to myself, I seem to
have been only like a child playing on the seashore, and diverting myself
in now and then finding a smoother pebble or a prettier shell than
ordinary, whilst the great ocean of truth lay all undiscovered before me.

—ISAAC NEWTON

THIS BOOK EXPLORES PATHS on the mathematical seashore. Paths are
the accumulated footprints of those who came before. There are
many paths to choose from — some leading to minor curiosities and
others leading to important mathematical goals. In this book, I intend to
point out a few paths that I believe are both curious and important — paths
with mainstream destinations.

I intend to show mathematics as a human endeavor, not a cold unap-
proachable monolithic perfection. The search for a useful, convincing, and
reliable understanding of number and space has had many successes, but
also a few false starts and wrong turns. Botticelli's famous painting *The
Birth of Venus* shows Venus born of the foam of the sea, not as an infant but
as a beautiful woman, divine in every detail. Mathematics, on the other
hand, did not achieve such instant perfection at birth. Her growth has
been long and tortuous, and perfection may be out of reach.

Part I of this book deals with the concept of number. We begin with
the curious method of the ancient Egyptians for representing fractions.
Fractions were a difficult concept for the ancients, and still are for today's
schoolchildren. Unguided, the mathematical pioneers discovered over
centuries what today's schoolchildren, guided by their teachers, learn in
weeks. The Egyptian method seems clumsy to us, yet we will see that it
provides some advantages in dividing five pies among seven people.

Part II is devoted to geometry. We will visit a fantastic universe called
Tubeland. The efforts of Tubelanders to understand their world is a reflection of the efforts of our scientists to understand ours. Question: What
geometric device was unknown in 1800, a promising innovation in 1900,
and a universal commonplace in mathematics, science, business, and everyday life in 2000? Answer: *Graphs.*

1

Part III is concerned with algebra, the language of mathematics. Solving equations was a passionate undertaking for five Italian mathematicians of the sixteenth century. For them, algebraic knowledge was booty of great value, the object of quarrels, conspiracies, insults, and fiery boasts. Later, we discover what a catalog of wallpaper ornaments has to do with algebra.

Part IV introduces the smoother pebble discovered by Newton and Leibniz: the calculus. The basic concepts are introduced by means of a six-minute automobile ride. Later, we witness the competition for the fastest roller coaster.

I hope that the reader gains from this book new meaning and new pleasure in mathematics.

Part I

Bridging the Gap

Science is the attempt to make the chaotic diversity of our
sense-experiments correspond to a logically uniform system of thought.

—ALBERT EINSTEIN (1879–1955)

1

Ancient Fractions

The Eye of Horus burning with fire before my eyes.

—THE BOOK OF THE DEAD, 1240 BCE
(translated by E.A. Wallis Budge)

POSITIVE WHOLE NUMBERS — the *natural numbers* — fill the fundamental human need for counting, but, additionally, a civilized society requires fractional numbers for the orderly division of land and goods — *artificial numbers* that fill in the gaps between the natural numbers.

Getting fractions right is the first slippery step in the mathematical education of many schoolchildren, a place where many fall. So it was also in the history of mathematics. The ancient Egyptians took a wrong turn. Only after thousands of years did others find the right path. This detour is now all but forgotten, and there is no danger that we will repeat this mistake. Since fractions were not easy for the Egyptians, we can be more understanding of the difficulties that our schoolchildren experience. Furthermore, Egyptian fractions are a source of curious problems, interesting in their own right.

The ancient Babylonians must be given high marks for their treatment of fractions. Babylonian fractions were quite similar to today's decimal fractions; however, the Babylonian system was based on the number 60 instead of 10. We still use Babylonian fractions when we use minutes and seconds to measure time and angles.

The German mathematician Leopold Kronecker (1823–91) said, "God created the whole numbers. All the rest is the work of man." There is essentially one way to understand the natural numbers. However, there are several different ways to define fractional numbers — also known as *rational numbers*. The fractions in current use — a numerator and denominator separated by a bar, for example, 5/7 — we call *common fractions*. This notation originated in India in the twelfth century and soon spread to Europe,

5

but the underlying concept — *ratios of commensurable magnitudes* — is from the ancient Greeks. However, common fractions are not the only way to conceive of fractions. In this chapter, we will see that the ancient Egyptians and Babylonians had different methods. In Chapter 2, we will see yet another method of defining fractional numbers.

The Egyptian Unit Fractions

The Rhind Papyrus, a scroll that measures 18 feet by 13 inches, is the most important source of information concerning ancient Egyptian mathematics. It was found in Thebes and purchased by Scottish Egyptologist Henry Rhind in 1858; it has been held by the British Museum since 1863. The scroll was written by the scribe Ahmes (1680?–1620? BCE), who states that he copied the material from older sources — 1850 BCE or earlier.

The scroll — consisting of two tables and 87 problems — is a textbook of ancient Egyptian mathematics. Some of the problems deal with areas and volumes; however, a considerable part of the scroll is concerned with the ancient Egyptian arithmetic of fractions. Despite obvious shortcomings, these curious methods persisted for thousands of years. In fact, we will see that Leonardo of Pisa (1175?–1230?)[1] made an important contribution to the theory of Egyptian *unit fractions*.

The ancient Egyptians devised a concept of fractions that seems strange — even bizarre — to us today. A fraction with numerator equal to 1 (e.g., 1/3, 1/7) is called a unit fraction. The Egyptians denoted unit fractions by placing the eye-shaped symbol $\bigcirc$ ("the eye of Horus") above a natural number to indicate its *reciprocal*. We approximate this notation by using, for example, $\overline{7}$ to represent 1/7.

The Egyptians had a special notation for 2/3, but all other fractions were represented as sums of distinct unit fractions. For example, for 5/7 they could have written

$$\frac{5}{7} = \overline{2} + \overline{7} + \overline{14} \tag{1.1}$$

We confirm this by the following computation:

$$\frac{1}{2} + \frac{1}{7} + \frac{1}{14} = \frac{7+2+1}{14} = \frac{10}{14} = \frac{5}{7} \tag{1.2}$$

Similar computations show that the fraction 5/7 can also be represented as

$$\frac{5}{7} = \overline{2} + \overline{5} + \overline{70} \tag{1.3}$$

or

$$\frac{5}{7} = \overline{3} + \overline{4} + \overline{8} + \overline{168} \tag{1.4}$$

It did not occur to the Egyptians to use *two* numbers, a numerator and denominator, to represent a fraction. When we write 5/7, and when we calculate as in equation (1.2) above, we are departing from the ancient Egyptian mode of thought.

Why did the Egyptians avoid repetitions of unit fractions? Why did they feel, for example, that $\overline{7} + \overline{7} + \overline{7} + \overline{7} + \overline{7}$ is unacceptable? One can only speculate, but perhaps they felt that it is not permissible to express a fraction as a sum of five unit fractions when three (as in equations (1.1) and (1.3)) are all that are needed.

The fact—show in equations (1.1), (1.3), and (1.4)—that 5/7 has more than one representation as a sum of unit fractions indicates a serious flaw in the Egyptian system for fractions. How is it possible that such an awkward system remained in use for thousands of years? There are several possible answers:

1. The system was adequate for simple needs.

2. The system was sanctioned by tradition.

3. The scribes who used the system had no wish to diminish their reputations for wizardry by simplifying the system.

4. It really does take thousands of years to get the bright idea that one can use *two* natural numbers—numerator and denominator—to specify a fraction.

Aside from the merit of the above speculations, there are certain real advantages in using Egyptian unit fractions for problems involving the division of goods. A fair method of division divides the whole into a number of pieces and specifies the pieces in each share. If we assume that the goods in question are *fungible*,[2] then the most important requirement for a method of fair division is that the total size of each share be identical regardless of the number and shape of the pieces forming each share. However, there are other considerations. For example, it adds to the *appearance* of fairness if the shares are identical—not only in aggregate size, but also in the number and shape of the pieces. Furthermore, the number of pieces should not be excessive. As shown by the following example, unit fractions can lead to a division of goods with certain advantages.

Example 1.1. Divide 5 pies among 7 people, Ada, Ben, Cal, Dot, Eli, Fay, and Gil, (a) using ordinary arithmetic, (b) using Egyptian unit fractions.

(a) **Two methods using ordinary arithmetic.**

Method 1:

1. Ada gets 5/7 of the first pie.
2. Ben gets 2/7 of the first pie and 3/7 of the second pie.
3. Cal gets 4/7 of the second pie and 1/7 of the third pie.
4. Dot gets 5/7 of the third pie.
5. Eli gets 1/7 of the third pie and 4/7 of the fourth pie.

6. Fay gets $3/7$ of the fourth pie and $2/7$ of the fifth pie.
7. Gil gets $5/7$ of the fifth pie.

Objection 1: Disagreements can arise because the shares contain different sized pieces.

Method 2: Divide each of the five pies into seven equal pieces. A share consist of five of these pieces.

Objection 2: Too many pieces in each share.

(b) **A method using Egyptian unit fractions.** In this method each share consists of just three pieces, and all the shares have the same appearance. Since, according to equation (1.1), $5/7 = 1/2 + 1/7 + 1/14$, we proceed as follows:

1. Give everyone half of a pie. This leaves a pie and a half to be distributed.
2. Cut the remaining whole pie in sevenths. Give each of the seven people one of these pieces. There remains half a pie.
3. Cut the half pie in seven equal pieces. Give each person one of these pieces.

The Egyptian method avoids the worst features of each of the two modern methods. It beats Method 1 on the grounds of Objection 1 and Method 2 on the grounds of Objection 2.

Egyptian arithmetic

Like schoolchildren of today, the Egyptians needed basic arithmetic as a background for computing with fractions. The Rhind Papyrus gives examples illustrating a complex collection of arithmetic techniques. We will consider some of the methods of multiplication and division — especially as they relate to fractions.[3]

Multiplication

The ancient Egyptians did not use our familiar methods of multiplication and division. The basic method of multiplication, which proceeds by successive doubling, is illustrated in Table 1.1(a). Successive doubling involves exactly the same arithmetic as *Russian peasant multiplication* (see Table 1.1(b)). Both methods convert one of the factors to the *binary system*, the arithmetic basis for the modern digital computer. In multiplying 13×14, the factor 13 is converted to binary ($13 = 8 + 4 + 1 = 1101_2$) — in Table 1.1(a) by starring certain items and in Table 1.1(b) by striking out certain items. The Russian peasant method is an improvement because it gives a mechanical process — an algorithm — for the binary conversion.

Table 1.1. Ancient and modern multiplication: $13 \times 14 = 182$.

(a)		(b)		(c)		(d)	(e)	
*1	14	13	14			$\begin{array}{r}14\\ \times\,13\\ \hline\end{array}$		
2	28	6	~~28~~	*1	14		1	14
*4	56	3	56	*2	28	42	*3	42
*8	112	1	112	*10	140	14	*10	140
13	**182**		**182**	13	**182**	**182**	13	**182**

(a) **Egyptian method (doubling only).**

   1. The top row consists of 1 and the second factor (14).
   2. Each successive row is obtained by doubling the preceding row.
   3. Place a star beside the numbers in the left column that sum to 13.
   4. The product 13×14 is the sum of the numbers in the right column in the starred rows.

(b) **Russian peasant multiplication.** This method of multiplication is essentially the same as (a):

1. The two numbers to be multiplied are entered as the top items in the two columns.
2. If the item in the first column is a 1, then we are finished. Otherwise, repeat the following two steps until a 1 appears in the first column: A. Append a number to the bottom of the first column equal to half of the number above it — ignoring any fractional amount. B. Append a number to the second column equal to double the number above it.
3. For each *even* number in the first column, strike out the adjacent number in the right column.
4. The product is equal to the sum of the numbers remaining in the second column. The striking out of elements of the second row that are adjacent to even numbers in the first column is equivalent to finding the representation of 13 in the binary system,[4] that is, $13_{10} = 1101_2$.

(c) **Egyptian method (shortcut).** In the Egyptian method, as shown in (a), the succeeding rows are obtained by doubling the preceding row. However, the method works equally if we multiply the elements of any of the preceding rows by any convenient natural number. To speed up the process the Egyptians sometimes appended a row obtained by multiplying the first row by 5 or 10. In this example, the third row is obtained by multiplying elements of the first row by 10.

(d) **Modern multiplication algorithm.**

(e) **Modern multiplication in the Egyptian format.** According to our place-value decimal system, the first factor 13 is an abbreviation for $10 + 3$. In the Egyptian format, we use 3 and 10 as multipliers to transform the first row into the second and third rows respectively. We obtain the product 182 as the sum of 42 and 140 both in the modern method (d) and the Egyptian method (e). (In the modern method, we write 140 instead of 14, but this 14 is shifted one place to the left — equivalent to multiplying 14 by 10.)

The Egyptians did not always proceed by doubling if there was an obvious shortcut. For example, in Table 1.1(c), they achieve $13 \times 14 = 182$ by multiplying the first row of the table by 1, 2, and 10. In this method of multiplication, we may multiply a row by any convenient number. This procedure, carried out suitably, results in exactly the same arithmetic operations as the familiar multiplication algorithm, shown in Table 1.1(d). In fact, Table 1.1(e) shows such a variant of the Egyptian method applied to 13×14. The multipliers, 3 and 10, come from the meaning of the decimal number 13:

$$13 = 1 \times 10^1 + 3 \times 10^0 = 10 + 3$$

The doubling table from the Rhind Papyrus

The Rhind Papyrus contains the curious Table 1.2 expressing fractions of the form $2/n$ as sums of distinct unit fractions. This table was an important Egyptian tool for computing with fractions; it lost its usefulness when new computational methods displaced unit fractions. (Centuries later, tables of logarithms suffered a similar fate. They were useful for arithmetic calculations—multiplication and exponentiation—until electronic calculators came into use in the 1970s.)

The left column of Table 1.2 does not contain fractions with even denominator $(2/2m)$ because the Egyptians readily computed $2/2m = \overline{m}$.

Table 1.2 was useful for adding Egyptian fractions. Since an Egyptian fraction is a sum of unit fractions with no duplication, the addition of two of these fractions might produce an illegal duplication of some unit fraction that could be resolved using Table 1.2. Example 1.2 shows how Table 1.2 could be used for arithmetic calculations.

Example 1.2. Use Table 1.2 to find the sum of the three fractions $\overline{5} + \overline{15}$, $\overline{10} + \overline{30}$, and $\overline{5} + \overline{25}$ (a) using standard modern arithmetic and (b) using Egyptian methods.

(a) **Modern method.** The problem is to add the following three fractions:

$$\frac{1}{5} + \frac{1}{15} = \frac{3+1}{15} = \frac{4}{15}$$

$$\frac{1}{10} + \frac{1}{30} = \frac{3+1}{30} = \frac{2}{15}$$

$$\frac{1}{5} + \frac{1}{25} = \frac{5+1}{25} = \frac{6}{25}$$

Using the common denominator 75, we find

$$\frac{4}{15} + \frac{2}{15} + \frac{6}{25} = \frac{4 \cdot 5 + 2 \cdot 5 + 6 \cdot 3}{75} = \frac{48}{75} = \frac{16}{25}$$

(b) **Egyptian method.** We must find a sum of

$$(\overline{5} + \overline{15}) + (\overline{10} + \overline{30}) + (\overline{5} + \overline{25})$$

Rearranging these terms, we have

$$(\overline{5} + \overline{5}) + \overline{10} + \overline{15} + \overline{25} + \overline{30}$$

Table 1.2. Doubling unit fractions (from the Rhind Papyrus).

Fractions of the form $2/n$ expressed as sums of distinct unit fractions	
$2/3 \ = \overline{2} + \overline{6}$	$2/53 \ = \overline{30} + \overline{318} + \overline{795}$
$2/5 \ = \overline{3} + \overline{15}$	$2/55 \ = \overline{30} + \overline{330}$
$2/7 \ = \overline{4} + \overline{28}$	$2/57 \ = \overline{38} + \overline{114}$
$2/9 \ = \overline{6} + \overline{18}$	$2/59 \ = \overline{36} + \overline{236} + \overline{531}$
$2/11 = \overline{6} + \overline{66}$	$2/61 \ = \overline{40} + \overline{244} + \overline{488} + \overline{610}$
$2/13 = \overline{8} + \overline{52} + \overline{104}$	$2/63 \ = \overline{42} + \overline{126}$
$2/15 = \overline{10} + \overline{30}$	$2/65 \ = \overline{39} + \overline{195}$
$2/17 = \overline{12} + \overline{51} + \overline{68}$	$2/67 \ = \overline{40} + \overline{335} + \overline{536}$
$2/19 = \overline{12} + \overline{76} + \overline{114}$	$2/69 \ = \overline{46} + \overline{138}$
$2/21 = \overline{14} + \overline{42}$	$2/71 \ = \overline{40} + \overline{568} + \overline{710}$
$2/23 = \overline{12} + \overline{276}$	$2/73 \ = \overline{60} + \overline{219} + \overline{292} + \overline{365}$
$2/25 = \overline{15} + \overline{75}$	$2/75 \ = \overline{50} + \overline{150}$
$2/27 = \overline{18} + \overline{52}$	$2/77 \ = \overline{44} + \overline{308}$
$2/29 = \overline{24} + \overline{58} + \overline{174} + \overline{232}$	$2/79 \ = \overline{60} + \overline{237} + \overline{316} + \overline{790}$
$2/31 = \overline{20} + \overline{124} + \overline{155}$	$2/81 \ = \overline{54} + \overline{162}$
$2/33 = \overline{22} + \overline{66}$	$2/83 \ = \overline{60} + \overline{332} + \overline{415} + \overline{498}$
$2/35 = \overline{30} + \overline{42}$	$2/85 \ = \overline{51} + \overline{255}$
$2/37 = \overline{24} + \overline{111} + \overline{296}$	$2/87 \ = \overline{58} + \overline{174}$
$2/39 = \overline{26} + \overline{78}$	$2/89 \ = \overline{60} + \overline{356} + \overline{534} + \overline{890}$
$2/41 = \overline{24} + \overline{246} + \overline{328}$	$2/91 \ = \overline{70} + \overline{130}$
$2/43 = \overline{42} + \overline{86} + \overline{129} + \overline{301}$	$2/93 \ = \overline{62} + \overline{186}$
$2/45 = \overline{30} + \overline{90}$	$2/95 \ = \overline{60} + \overline{380} + \overline{570}$
$2/47 = \overline{30} + \overline{141} + \overline{470}$	$2/97 \ = \overline{56} + \overline{679} + \overline{776}$
$2/49 = \overline{28} + \overline{196}$	$2/99 \ = \overline{66} + \overline{198}$
$2/51 = \overline{34} + \overline{102}$	$2/101 = \overline{101} + \overline{202} + \overline{303} + \overline{606}$

From Table 1.2, we see that $\overline{5} + \overline{5}$ ($= 2/5$) can be replaced by $\overline{3} + \overline{15}$. Rearranging again, we obtain

$$\overline{3} + \overline{10} + (\overline{15} + \overline{15}) + \overline{25} + \overline{30}$$

According to the table, $\overline{15} + \overline{15}$ ($= 2/15$) can be replaced by $\overline{10} + \overline{30}$. Rearranging again, we have

$$\overline{3} + (\overline{10} + \overline{10}) + \overline{25} + (\overline{30} + \overline{30})$$

Although neither $\overline{10} + \overline{10}$ ($= 2/10$) nor $\overline{30} + \overline{30}$ ($= 2/30$) appear in the table, the ancient scribes easily recognized that these are equal to $\overline{5}$ and $\overline{15}$, respectively, obtaining the final result

$$\overline{3} + \overline{5} + \overline{15} + \overline{25}$$

In modern notation, this is equal to

$$\frac{1}{3} + \frac{1}{5} + \frac{1}{15} + \frac{1}{25} = \frac{25 + 15 + 5 + 3}{75} = \frac{48}{75} = \frac{16}{25}$$

This agrees with the our calculation in (a).

Table 1.2 follows certain regular patterns.[5] For example, the denominators divisible by 3 follow the pattern

$$\frac{2}{3k} = \overline{2k} + \overline{6k}$$

The Egyptians do not tell us what patterns they used. Their use of patterns is based on the evidence of calculations as in Table 1.2.

Division

To divide a natural number m by a natural number n, the ancient Egyptians multiplied m by $1/n$ using the doubling algorithm of Table 1.1(a). The doubling process makes it necessary to use Table 1.2 to resolve doubled unit fractions into standard Egyptian fractions.

For example, Table 1.3(a) shows how this algorithm is applied to compute 5 divided by 7. We have already seen $5 \div 7$ expressed as a sum of unit fractions in equation (1.1), but in Table 1.3, we see how the Egyptians could have derived this relationship using their multiplication algorithm.

Table 1.3(b) shows a method of calculating $28 \div 13$ by successive doubling — possibly as proposed by an apprentice scribe. This example is included not as a recommended method of computation but as an illustration of the method of successive doubling and the use of Table 1.2. A master scribe would probably first convert the improper fraction to a proper one ($28 \div 13 = 2\,2/13$) and then use the relation $2 \cdot \overline{13} = \overline{8} + \overline{52} + \overline{104}$ from

Table 1.3. Calculation of (a) $5 \div 7$ and (b) $28 \div 13$ by the method of successive doubling. Items in square brackets are obtained from Table 1.2.

(a)

Derivation of $5 \div 7 = \overline{2} + \overline{7} + \overline{14}$	
$*1$	$\overline{7}$
2	$2 \cdot \overline{7} = [\overline{4} + \overline{28}]$
$*4$	$\overline{2} + \overline{14}$
$5 = 1 + 4$	$\overline{7} + \overline{2} + \overline{14}$

(b)

Derivation of $28 \div 13 = 2 + 2 \cdot \overline{13} = 2 + [\overline{8} + \overline{52} + \overline{104}]$	
1	$\overline{13}$
2	$2 \cdot \overline{13} = [\overline{8} + \overline{52} + \overline{104}]$
$*4$	$2 \cdot \overline{8} + 2 \cdot \overline{52} + 2 \cdot \overline{104} = \overline{4} + \overline{26} + \overline{52}$
$*8$	$2 \cdot \overline{4} + 2 \cdot \overline{26} + 2 \cdot \overline{52} = \overline{2} + \overline{13} + \overline{26}$
$*16$	$2 \cdot \overline{2} + 2 \cdot \overline{13} + 2 \cdot \overline{26} = 1 + [\overline{8} + \overline{52} + \overline{104}] + \overline{13}$
28	$(\overline{4} + \overline{26} + \overline{52}) + (\overline{2} + \overline{13} + \overline{26}) + (1 + \overline{8} + \overline{13} + \overline{52} + \overline{104})$
	$= 1 + \overline{2} + \overline{4} + \overline{8} + 2 \cdot \overline{13} + 2 \cdot \overline{26} + 2 \cdot \overline{52} + \overline{104}$
	$= 1 + \overline{2} + \overline{4} + \overline{8} + [\overline{8} + \overline{52} + \overline{104}] + \overline{13} + \overline{26} + \overline{104}$
	$= 1 + (\overline{2} + \overline{4} + \overline{8} + \overline{8}) + \overline{13} + \overline{26} + \overline{52} + 2 \cdot \overline{104}$
	$= 1 + 1 + \overline{13} + \overline{26} + 2 \cdot \overline{52}$
	$= 2 + \overline{13} + \overline{26} + \overline{26}$
	$= 2 + 2 \cdot \overline{13} = 2 + [\overline{8} + \overline{52} + \overline{104}]$

Table 1.2. Since the apprentice applied the method of successive doubling without errors, he arrived at the same result after a much more complex calculation.

We have seen that a fraction can have more than one representation as a sum of unit fractions. For example,

$$\frac{5}{7} = \overline{2} + \overline{7} + \overline{14} = \overline{2} + \overline{5} + \overline{70} = \overline{3} + \overline{4} + \overline{8} + \overline{168}$$

But do we know that there is always at least one such representation? The next section answers the following fundamental question.

Question 1.1. Can every proper fraction be expressed as a sum of distinct unit fractions?

The greedy algorithm

The answer to Question 1.1 is Yes. This was first shown by Leonardo of Pisa (Fibonacci) in 1202 in his work *Liber Abaci*. He showed that every proper fraction can be expressed as a sum of distinct unit fractions by a method now called the *greedy algorithm*. The following example illustrates the algorithm and shows why it is called greedy.

Example 1.3. Express the fraction 7/213 as a sum of unit fractions.

Solution. Start by finding the largest unit fraction not exceeding 7/213. (Since we are greedy, we look for the *largest*.) Because 213/7 = 30.4 ... , it follows that 7/213 is between 1/31 and 1/30, and the unit fraction that we seek is 1/31. Now subtract 1/31 from 2/213, obtaining

$$\frac{7}{213} - \frac{1}{31} = \frac{31 \times 7 - 213}{213 \times 31} = \frac{4}{6,603} \tag{1.5}$$

Greedy again, we seek the largest unit fraction not exceeding 4/6,603. Because 6,603/4 = 1,650.75, it follows that the unit fraction that we seek is 1/1651. We have

$$\frac{4}{6,603} - \frac{1}{1,651} = \frac{4 \times 1,651 - 6,603}{6,603 \times 1,651} = \frac{6,604 - 6,603}{10,901,553} = \frac{1}{10,901,553} \tag{1.6}$$

We have found the following representation of 7/213:

$$\frac{7}{213} = \overline{31} + \overline{1,651} + \overline{10,901,553}$$

We have shown that the greedy method succeeds in this particular case. How can we show that it *always* succeeds? It is daunting that the method generates such big denominators so quickly. Nevertheless, we see in this example the numerators become smaller: **7** → **4** → **1**. These numerators are shown in boldface (**7** and **4** in equation (1.5); **4** and **1** in equation (1.6)). This crucial clue enables us to show that the greedy method always works. The numerators are natural numbers; if they become successively smaller, they must eventually reach 1, the *smallest* natural number. Attention to a few algebraic details will repay the reader with the exhilaration of understanding the elegant proof of the following proposition.

Proposition 1.1. *Let $r = p/q$ be a proper fraction (i.e., $p/q < 1$). Then either (a) r is already a unit fraction, or (b) one iteration of the greedy algorithm yields a fraction with a numerator that is a natural number less than p.*

Proof. Suppose that r is not a unit fraction. Then r must be between two successive unit fractions. That is, there must be a natural number t (greater than 1) such that r is between 1/t and 1/(t−1). This fact is expressed algebraically as follows:

$$\frac{1}{t} < \frac{p}{q} < \frac{1}{t-1} \tag{1.7}$$

Note that the second inequality in this chain implies $pt - p < q$; adding p to, and subtracting q from, both sides of this inequality, we obtain

$$pt - q < p \qquad (1.8)$$

Now recall what we must prove. We must show that $p/q - 1/t$ is equal to a fraction with a denominator that is a natural number less than p. Using the common denominator qt and the first inequality of (1.7), we obtain

$$\frac{p}{q} - \frac{1}{t} = \frac{pt - q}{qt} > 0$$

But now we are finished because inequality (1.8) tells us what we want to know, that the (positive) numerator $pt - q$ is less than the numerator p.

$\square$

Now we can answer Question 1.1 affirmatively. Starting with an arbitrary proper fraction, successive applications of the greedy algorithm must eventually yield a unit fraction. The original fraction is equal to the sum of the unit fractions obtained by finitely many applications of the greedy algorithm.

A given proper fraction has a *unique* greedy representation as a sum of distinct unit fractions. The representations in Table 1.2 are not all greedy. In fact, the greedy representation of $2/(2n-1), n = 2, 3, \ldots$ consists of the sum of just two unit fractions, as shown by the following formula:

$$\frac{2}{2n-1} = \frac{(2n-1)+1}{n(2n-1)} = \overline{n} + \overline{n(2n-1)}$$

In Table 1.2, only the representations of $2/3$, $2/5$, $2/7$, $2/11$, and $2/23$ can be obtained using this formula; all the others are not greedy.

In general, an iterative algorithm is called *greedy* if it seeks to maximize the partial outcome of each step. Some greedy algorithms succeed (e.g., Proposition 1.1), but others fail. In chess (and in life) it is unwise to capture a pawn while losing sight of other goals.

The Babylonians and the Sexagesimal System

Another method of representing fractions is due to the ancient Babylonians. Their system of representing numbers is similar to our decimal system; however, they used 60 instead of 10 as a base for their number system—a *sexagesimal system* instead of a decimal system. We represent sexagesimal numbers in the manner of the following example:

$$11, 0, 21; 12, 45 = 11 \cdot 60^2 + 0 \cdot 60 + 21 + \frac{12}{60} + \frac{45}{60^2}$$

We use commas to separate the digits and a semicolon (instead of a decimal point) to separate the *integer* part from the fractional part. This form is convenient for us, but by using it we are giving the Babylonians a bit too much credit because they did not have a zero.

We have inherited the Babylonian sexagesimal system for measuring time—60 seconds in a minute and 60 minutes in an hour. In measuring angles, seconds and minutes are also used for sexagesimal fractions of a degree. The sexagesimal system was used by the ancient Greeks, and it was used in Europe as late as the sixteenth century when the decimal system was introduced.

Sexagesimal fractions

In the decimal system, 1/3 is equal to the awkward infinite decimal fraction 0.333... . The sexagesimal system scores an advantage here because 1/3 is represented by the simpler terminating fraction 0; 20. The *prime factorization* of 60, the base of the sexagesimal system, is $2^2 \cdot 3 \cdot 5$. Any natural number that is not divisible by any *prime numbers* other than 2, 3, and 5 is called a *regular* sexagesimal number. They are the only natural numbers with reciprocals that can be represented as *terminating* sexagesimal fractions. For example, the number 54 is a regular sexagesimal number because it has the prime factorization $54 = 2 \cdot 3^3$, and its reciprocal, 1/54, has the terminating sexagesimal representation 0; 1, 6, 40. We confirm this representation by the following calculation:

$$0; 1, 6, 40 = \frac{1}{60} + \frac{6}{60^2} + \frac{40}{60^3} = \frac{60^2 + 6 \cdot 60 + 40}{60^3} = \frac{4000}{216000} = \frac{1}{54}$$

On the other hand, the sexagesimal system has the disadvantage of having a much larger multiplication table than the decimal system. For the decimal system, it suffices to memorize 45 products in the multiplication table up to 9×9; however, to do sexagesimal arithmetic we need to know all 1,770 entries of the multiplication table up to 59×59. However, if one masters the sexagesimal multiplication table, more rapid arithmetic computations are possible than with the decimal system.

Babylonian cuneiform tablets contain many tables of numerical calculations. There are tables that imply a knowledge of the Pythagorean Theorem long before Pythagoras. To facilitate calculations with fractions, the Babylonians used a table (e.g., Table 1) of reciprocals of ordinary sexagesimal numbers. An important use of Table 1 is to replace division by a natural number n to multiplication by 1/n. Multiplication was achieved by a process equivalent to the algorithm now known as Russian peasant multiplication.

Table 1.4. The information in this table is found in a number of Babylonian cuneiform tablets. For convenience, this table uses modern notation.

Reciprocals of regular sexagesimal integers from 1 to 81			
1/2 = 0;30	1/16 = 0;3,45	1/45	= 0;1,20
1/3 = 0;20	1/18 = 0;3,20	1/48	= 0;1,15
1/4 = 0;15	1/20 = 0;3	1/50	= 0;1,12
1/5 = 0;12	1/24 = 0;2,30	1/54	= 0;1,6,40
1/6 = 0;10	1/25 = 0;2,24	1/60 = 1/1,0;	= 0;1
1/8 = 0;7,30	1/27 = 0;2,13,20	1/64 = 1/1,4;	= 0;0,56,15
1/9 = 0;6,40	1/30 = 0;2	1/72 = 1/1,12;	= 0;0,50
1/10 = 0;6	1/32 = 0;1,52,30	1/75 = 1/1,15;	= 0;0,48
1/12 = 0;5	1/36 = 0;1,40	1/80 = 1/1,20;	= 0;0,45
1/15 = 0;4	1/40 = 0;1,30	1/81 = 1/1,21;	= 0;0,44,26,40

The Athenian Greek mathematicians of the fifth century BCE — the Periclean "golden age" — did not inherit the Babylonian interest in numerical calculation and did not use the sexagesimal system. They persisted in using the awkward Egyptian unit fractions. Later, however, the Alexandrian Greek astronomers, particularly Claudius Ptolemy (85?–165? CE), used sexagesimal numbers.

Leonardo of Pisa (Fibonacci), who first proved the greedy algorithm (Proposition 1.1) for Egyptian unit fractions, was also acquainted with decimal and sexagesimal numbers. He used sexagesimal numbers to give a solution, accurate to the equivalent of nine decimal places, of a certain cubic equation.

Egyptian unit fractions have long ago ceased to be a tool for serious computation. Today they are merely a source of curious problems.

On the other hand, we are the direct beneficiaries of the Babylonian place-value sexagesimal system, although we happen to use base 10 instead of 60. The system merely needed fine-tuning — the proper usage of zero.

The ancient Greek mathematicians were more interested in theory than computation. However, they introduced the idea of mathematical proof — the measuring stick by which all mathematics is now validated. In the next chapter, we will see that Greek ideas of ratio and proportion brought about a deeper new understanding of the number system.

2

Greek Gifts

Socrates: *What do you say of him, Meno? Were not all these answers given out of his own head?*
Meno: *Yes, they were all his own.*
Socrates: *And yet, as we were just now saying, he did not know?*
Meno: *True.*
Socrates: *But still he had in him those notions of his — had he not?*
Meno: *Yes.*
Socrates: *Then he who does not know may still have true notions of that which he does not know?*
Meno: *He has.*

—PLATO (427?–347? BCE), Meno (translated by Benjamin Jowett)

T HE ANCIENT GREEKS EMBARKED HUMANITY on the scientific voyage of discovery that continues to the present day. The greatest of the Greek mathematical gifts to us from antiquity was the notion of *proof*. In this chapter, we look at certain fundamental accomplishments of rigorous Greek mathematical thought — two alternate developments of the theory of ratio and proportion, subtle methods of filling in the gaps between the whole numbers. Thereby the Greeks carried forward the development of the number system.

The Greeks looked beyond the practical needs of society for counting and measuring. Their findings come to us largely through the *Elements* of Euclid (fl. 295? BCE). We are fortunate to have such a record of ancient Greek mathematics; however, the beginnings of Greek mathematics are more shadowy. Thales of Miletus (625?–546? BCE) and Pythagoras of Samos (580?–500? BCE) were the first Greek mathematicians. Miletus was a Greek coastal city of Asia Minor (now Turkey), and Samos is a Greek island — both are on the Aegean Sea. Both men are said to have brought back knowledge from travels to Mesopotamia and Egypt. Regrettably, this knowledge does not seem to have included the superior Babylonian sexagesimal system and its place-value system of representing numbers. However, they more than made up for this lack by originating geometry as a deductive science — a uniquely Greek contribution to mathematics with no counterpart in Mesopotamia or Egypt.

Thales is said to have been the first to conceive of geometry as a chain of logical deduction — from axioms to theorems. He is said to have given proofs of several theorems, but we are told this only by commentators who came hundreds of years later.

Neither Thales nor Pythagoras left us a written account of their discoveries. However, Pythagoras left a cadre of disciples who carried his teachings forward. In fact, Pythagoras founded a secret society at Croton on the southeast coast of Italy — then called Magna Graecia.

The Pythagoreans were at the same time cultists and scientists. On the one hand, they were a secret brotherhood that found mystical significance in numbers. On the other hand, they discovered mathematical truths and promulgated the concept of *mathematical proof*. They studied especially the properties of the whole numbers — even and odd, divisibility, prime numbers, and so on. The Greeks called this branch of mathematics *arithmetic*, and we now call it *number theory*.

"All is number" was the motto of the Pythagoreans. Nevertheless, the Greek mathematicians who followed found much more to say about geometry than number. The reluctance to integrate geometric and numerical magnitudes was an impediment to the progress of Greek mathematics. In this chapter, we will see that the concepts of ratio and proportion carried them to the brink of reconciling these concepts.

The later Greek mathematicians were philosopher-scientists seeking the truth and imparting it to others. Their greatest contributions to mathematics were in geometry. They made a sharp distinction between geometry and arithmetic. We will look especially at the distinction that they made between arithmetic magnitudes (numbers) and geometric magnitudes (lengths, areas, and volumes). This distinction is illustrated in the following quotation from *Posterior Analytics*[1] by Aristotle (384–322 BCE):

> The axioms which are premises of demonstration may be identical in two or more sciences: but in the case of two different genera such as arithmetic and geometry you cannot apply arithmetical demonstration to the properties of magnitudes unless the magnitudes in question are numbers.

The distinction between geometric and arithmetic magnitudes seems artificial today because the modern *real number system* does not distinguish between geometric and arithmetic magnitudes. Nevertheless, we preserve a vestige of this obsolete dichotomy in mathematical terms that we have inherited from the Greeks, such as *geometric* and *arithmetic progressions*.

Today, the real number system underlies our mathematics education — albeit only implicitly, since real numbers are studied explicitly only in a few advanced college-level courses. Unlike the ancient Greeks, when we approach a problem in geometry, we use numbers freely without separating geometric and arithmetic magnitudes. In other words, our geometry is completely *arithmetized*.

The German philosopher Hegel (1770–1831) declared that history progresses in cycles of *thesis, antithesis,* and *synthesis* — from innocence, to conflict, and finally to resolution. We can see this in the history of the concept of number. In the preceding chapter, we have seen the practical, unsophisticated number concepts of the Egyptians and Babylonians (thesis). In this chapter, we will see how the Greeks introduced new concepts (antithesis), and foreshadowed the modern concept of real number (synthesis).

The Heresy

It is said that the Pythagoreans punished those who divulged their secrets. This may be a calumny promulgated by outsiders suspicious of this secret brotherhood. Truth or legend, it is said that Hippasus of Metapontum (400? BCE) was drowned at sea by the Pythagoreans for divulging the proof that the side and diagonal of a square are incommensurable. Later, we discuss this result in detail. This proof is an important mathematical milestone for three reasons:

1. It is a proof of unexcelled logical beauty — a model of mathematical elegance.

2. It defines a major concern of ancient Greek mathematics. It is controversial whether the Greeks, themselves, perceived the existence of incommensurables as a crisis in the foundations of mathematics. However, in retrospect, we can say that, even if the Greeks did not see it, there was a turning point; and it is of interest how the Greeks found a resolution.

3. It is one of the very earliest instances of a *mathematical!proof*. The Greeks were the first to understand that mathematical truth could be established, not by authority, but by a self-contained convincing argument — that anyone with the patience to follow a logical discourse can see the truth. This point of view is illustrated in the above epigraph from Plato's dialog *Meno* in which Socrates has just proved a theorem of geometry to an uneducated slave boy.

In fact, the special case of the Pythagorean theorem that Socrates presented is relevant to our discussion. The large square in Figure 2.1 consists of eight congruent isosceles right triangles. The area of the square bounded by the four diagonal lines is twice the area of the small shaded square because the diagonal square consists of four triangles and the shaded square consists of two triangles. The side of the diagonal square is the diagonal of the shaded square.

Thus, if a and c denote the side and diagonal, respectively, of the shaded square, we have

$$2a^2 = c^2 \qquad (2.1)$$

The numerical values of a and c depend on the unit of measurement—for example, feet, millimeters, or angstroms. One might think that both the side and diagonal of the square could be integer multiples of some sufficiently small unit. The alleged crime of Hippasus consisted in revealing the "logical scandal" —*however small the unit of measurement, a and c cannot both be integers.*

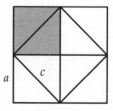

Figure 2.1.

Proposition 2.1. *The lengths of the side and diagonal of a square cannot both be integers.*

Proof. Suppose that a and c are integers satisfying equation (2.1). We begin by canceling any common factor. Suppose that k is the greatest common factor between a and c. Then there are integers A and C such that $a = kA$ and $c = kC$. Hence, from equation (2.1), we have $2k^2A^2 = k^2C^2$, which implies

$$2A^2 = C^2 \qquad (2.2)$$

where A and C have no common factor. Equation (2.2) implies that C is even. The square of an even number must be divisible by 4; therefore, the right side of equation (2.2) is divisible by 4.

Since C is even and there is no common factor between C and A, A must be odd. The square of an odd number is odd. It follows that the left side of equation (2.2) is divisible by 2 but not by 4.

We are finished. It is not possible that the right side of equation (2.2) is divisible by 4 and the left side is not. Our assumption that equation (2.1) holds must be false. ☐

In modern terms, Proposition 2.1 is equivalent to the assertion that $\sqrt{2}$ is an *irrational number.*

In geometry, two line segments $\mathcal{I}$ and $\mathcal{J}$ are called *commensurable* —that is, they have a "common measure" —if there exists a unit segment $\mathcal{U}$ such that both $\mathcal{I}$ and $\mathcal{J}$ can be covered by an integral number of nonoverlapping copies of $\mathcal{U}$. In Figure 2.2, the intervals $\mathcal{I}$ and $\mathcal{J}$ are covered by, respectively, four and five nonoverlapping copies of the unit interval $\mathcal{U}$. In this case, the *ratio* of $\mathcal{I}$ to $\mathcal{J}$ is 4 : 5. We can interpret this ratio as the *fraction* 4/5.

Proposition 2.1 asserts that the side and diagonal of a square are incommensurable. That is, the ratio of the side to the diagonal is not equal

$$\mathcal{U} \qquad \mathcal{I} \qquad \mathcal{J}$$

Figure 2.2. The intervals $\mathcal{I}$ and $\mathcal{J}$ are commensurable with respect to the unit interval $\mathcal{U}$. In fact, four nonoverlapping copies of $\mathcal{U}$ cover $\mathcal{I}$, and five of them cover $\mathcal{J}$, that is, $\mathcal{I} = 4\mathcal{U}$ and $\mathcal{J} = 5\mathcal{U}$. Hence, the ratio of the lengths of these two intervals is 4 : 5.

to a ratio of two natural numbers. Therefore, numerical magnitudes are not sufficient for describing ratios of geometric magnitudes. A theory of geometric ratios is needed.

What could be the motive for the alleged murder of Hippasus? It is said that he committed a grave sacrilege by denying the deeply held belief of the Pythagoreans that "number is all." Indeed, Proposition 2.1 seems to say that numbers—more specifically, the *natural* numbers—are not even powerful enough to resolve a simple geometric matter concerning the diagonal of a square. However, as we will see in discussion of Example 2.2 on page 29, the natural numbers *are* sufficient to explain this seeming paradox—through a process called *anthyphairesis*—a Greek word meaning *back-and-forth subtraction*. It is not surprising that the Pythagoreans failed to understand this mitigation of Hippasus' crime. Indeed, anthyphairesis is a subtle and beautiful concept that continues to unfold in the present time—especially in the *theory of continued fractions*.[2]

Magnitudes, Ratio, and Proportion

The ancient Greeks may or may not have perceived that the existence of incommensurable magnitudes created a crisis in the foundations of their mathematics. At any rate, they resolved this "logical scandal" by making a distinction between geometric and arithmetic magnitudes and by developing a theory of ratio and proportion. In ordinary usage, ratio and proportion are sometimes used interchangeably, but here we will make a more careful distinction. Specifically, an equality of ratios is called a *proportion*. The Greek concepts of ratio and proportion are close to what we now call *real numbers*.

Ratio and proportion are concerned with *magnitudes*. To understand the Greek point of view, we must suppress our modern conviction that all magnitudes are numbers. For the Greeks, there were several incompatible classes of magnitudes. The Greeks did not have a concept of zero, negative, or infinite magnitudes.

Book V of Euclid's *Elements*,[3] attributed to the Greek mathematician Eudoxus (400?–347? BCE), contains the following passage concerning magnitude and ratio.

Definition 3. A ratio is a sort of relation in respect of size between two magnitudes of the same kind.

Definition 4. Magnitudes are said to have a ratio to one another which can, when multiplied, exceed one another.

The above definitions set the stage, but they do not enable us to understand completely what is meant by ratio—beyond the idea that a ratio is something that depends on two magnitudes. These definitions show the properties of magnitudes that the Greeks wished to emphasize in their theory of ratio and proportion.

Definition 3 speaks of "magnitudes *of the same kind*." Arithmetic magnitudes are *numbers*—more specifically, natural numbers; geometric magnitudes can be *lengths, areas,* or *volumes.* These four kinds of magnitudes do not exhaust all possibilities; for example, Archimedes makes use of a geometric magnitude called *moment.* Two magnitudes are "of the same kind" if both are numbers, both are lengths, both are areas, or both are volumes, and so on.

Magnitudes are ordered: Given two magnitudes of the same kind, either they are equal or one is larger than the other. Furthermore, certain arithmetic operations of magnitudes are implied by the ancient Greek usage. This arithmetic is evident for numerical magnitudes, but requires some explanation for geometric magnitudes:

1. **The addition of two like magnitudes.** The sum of two geometric magnitudes is the magnitude of a geometric figure consisting of the two underlying figures side by side.
2. **The multiplication of a magnitude by a natural number.** Definition 4 implies that each multiple of a magnitude—double, triple, or any integral multiple—is also a magnitude of the same kind. In other words, if A is a geometric magnitude and n is a natural number, it makes sense to speak of the magnitude nA, the n-fold multiple of A. For example, $2A$ is the magnitude of a figure consisting of two copies of a figure of magnitude A.
3. **The subtraction of the smaller from the larger of two like magnitudes.** If the geometric magnitude A is larger than B, then $A - B$ is the magnitude of a geometric object of magnitude A from which an object of magnitude B has been removed.

We can now restate Definition 4 in a form that is known as the *Axiom of Archimedes.*

Axiom 2.1 (Axiom of Archimedes). *If A is a magnitude of the same kind as B, and A exceeds B ($A > B$), then a sufficiently large multiple of B exceeds A; that is, there exists a natural number n such that the n-fold multiple of B exceeds A ($nB > A$).*

The ratio of two magnitudes, $\mathcal{A}$ and $\mathcal{B}$, is written $\mathcal{A} : \mathcal{B}$. In modern terms, $\mathcal{A} : \mathcal{B}$ is like the quotient $\mathcal{A}/\mathcal{B}$. There are two possible methods for giving meaning to ratios in the ancient context:

1. **The method of Eudoxus.** Define what it means for one ratio to be equal to or greater than another ratio.[4] As we will see in the next section, this can be done even if we have not defined what a ratio $\mathcal{A} : \mathcal{B}$ actually *is*. Ratios can be left undefined just as lines and points in geometry are undefined.

2. **Anthyphairesis.** Define a ratio $\mathcal{A} : \mathcal{B}$ in terms of the natural numbers even when $\mathcal{A}$ and $\mathcal{B}$ are not numerical magnitudes. Further below, we will see how this can be done.

There is no inconsistency between method 1 and method 2. Book V of Euclid's *Elements* develops method 1. The role of method 2 can be ascertained only in part by a literal reading of Euclid. However, some scholars are confident, based on indirect evidence, that the Greeks used method 2 much more than the literal written record might indicate.[5]

We now look at methods 1 and 2 in more detail.

Method 1 — proportion according to Eudoxus

This section discusses Eudoxus's theory of proportion, presented in Book V of Euclid's *Elements*. Eudoxus's theory of proportion deals with both numerical and geometric magnitudes. A proportion is a relation of equality between two ratios. The traditional notation for a proportion between two ratios is $\mathcal{A} : \mathcal{B} :: \mathcal{R} : \mathcal{S}$, but this formula has the same meaning as $\mathcal{A} : \mathcal{B} = \mathcal{R} : \mathcal{S}$. Magnitudes $\mathcal{A}$ and $\mathcal{S}$ are called the *extremes*, and $\mathcal{B}$ and $\mathcal{R}$ are called the *means* of the proportion $\mathcal{A} : \mathcal{B} :: \mathcal{R} : \mathcal{S}$. Eudoxus's theory deals with magnitudes in general, but to make the following discussion less abstract, magnitudes are interpreted as line segments.

Definition 2.1 (Eudoxus). Let $\mathcal{I}$, $\mathcal{J}$, $\mathcal{K}$, and $\mathcal{L}$ be line segments. We say that the ratios $\mathcal{I} : \mathcal{J}$ and $\mathcal{K} : \mathcal{L}$ are equal if for every pair of natural numbers m and n, exactly one of the following three possibilities is true:

1. $m\mathcal{I} < n\mathcal{J}$ and $m\mathcal{K} < n\mathcal{L}$
2. $m\mathcal{I} = n\mathcal{J}$ and $m\mathcal{K} = n\mathcal{L}$
3. $m\mathcal{I} > n\mathcal{J}$ and $m\mathcal{K} > n\mathcal{L}$

Definition 2.1 is stated in Book V of Euclid's *Elements* as follows:

> **Definition 5.** Magnitudes are said to be in the same ratio, the first to the second and the third to the fourth, when, if any *equimultiples* whatever are taken of the first and third, and any equimultiples whatever of the second and fourth, the former equimultiples alike exceed, are alike equal to, or alike fall short

of, the latter equimultiples respectively taken in corresponding order.

Definition 6. Let magnitudes which have the same ratio be called proportional.

To understand the meaning of Definition 2.1, let us adopt a modern point of view for the moment. Suppose that $\mathcal{I}$ and $\mathcal{J}$ are the diagonal and side, respectively, of a particular square, and that $\mathcal{K}$ and $\mathcal{L}$ are the diagonal and side of a different square, as shown in Figure 2.3. From our modern point of view, we know that the ratios $\mathcal{I} : \mathcal{J}$ and $\mathcal{K} : \mathcal{L}$ are both equal to $\sqrt{2}$, and, hence, the proportion $\mathcal{I} : \mathcal{J} :: \mathcal{K} : \mathcal{L}$ is true. Furthermore, we can verify that $\sqrt{2}$ is between 1.41 and 1.42 because

$$1.41^2 = 1.9881 < 2 < 1.42^2 = 2.0164$$

From the relation

$$\mathcal{I} : \mathcal{J} = \mathcal{K} : \mathcal{L} = \sqrt{2} < 1.42 = \frac{142}{100}$$

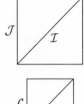

we have $100\mathcal{I} < 142\mathcal{J}$ and $100\mathcal{K} < 141\mathcal{L}$. In other words, putting $m = 100$ and $n = 142$, possibility 1 holds in Definition 2.1. On the other hand, since

$$\mathcal{I} : \mathcal{J} = \mathcal{K} : \mathcal{L} = \sqrt{2} > 1.41 = \frac{141}{100}$$

we have $100\mathcal{I} > 141\mathcal{J}$ and $100\mathcal{K} > 141\mathcal{L}$. In other words, putting $m = 100$ and $n = 141$,

Figure 2.3.

possibility 3 holds in Definition 2.1. (For this example, there is no choice of m and n such that possibility 2 holds because, as we have seen from Proposition 2.1, $\sqrt{2}$ is an irrational number.)

Eudoxus's definition of proportion is essentially the same as the modern definition of real numbers from Richard Dedekind (1831–1916).[6] Both Eudoxus and Dedekind deal with the following sort of question: *How can we explain the real number $\sqrt{2}$ to a stubborn skeptic who insists that he knows only the natural numbers?* We cannot answer this question by completing the statement, "$\sqrt{2}$ is a *number* such that ...." Such an answer is circular because the only numbers that our skeptic is willing to accept are the *natural numbers*. The answer of Eudoxus and Dedekind, implicit in Definition 2.1, is that each pair of natural numbers m, n satisfies either $2n^2 < m^2$ or $2n^2 \geq m^2$, and this dichotomy of pairs of natural numbers *is* the real number $\sqrt{2}$. The real numbers in general can be defined in this manner — as dichotomies of the set of all pairs of natural numbers. The benefit of defining real numbers in this seemingly bizarre way is that it establishes that real numbers actually exist, provided that we accept the existence of the natural numbers.

Method 2 — Attributed to Theaetetus

The mathematical writings of Theaetetus (415?–369? BCE) have not sur-
vived, but scholars believe that Books X and XIII of Euclid's *Elements* are
a description of Theaetetus's work. It is unfortunate that, at best, we have
only a second-hand account that no doubt reflects the interest and understanding of Euclid.[7] Scholars agree that Theaetetus made fundamental contributions to the theory of proportion and incommensurables. Van
der Waerden (1975, p. 176) argues that certain propositions concerning ratios in Book X cannot easily be derived from Eudoxus's theory, and that,
therefore, Theaetetus must have used a different definition of ratio using
a method called *anthyphairesis*; we will call it *back-and-forth subtraction*, or
simply BAFS.

Starting from Axiom 2.1, the *Axiom of Archimedes*, we see that, given
a greater magnitude $\mathcal{A}$ and smaller magnitude $\mathcal{B}$, there must be *largest
multiple* of $\mathcal{B}$ that does not exceed $\mathcal{A}$. We state this more precisely in the
following proposition, which is currently known as the *division algorithm*.[8]

Proposition 2.2 (division algorithm). *Suppose that $\mathcal{A}$ is a magnitude of the
same kind as $\mathcal{B}$, and that $\mathcal{A}$ exceeds $\mathcal{B}$ ($\mathcal{A} > \mathcal{B}$). Then there is a largest multiple
of $\mathcal{B}$ that does not exceed $\mathcal{A}$. That is, there exists a natural number q, called the
quotient,* such that $q\mathcal{B} \leq \mathcal{A} < (q+1)\mathcal{B}$. If $q\mathcal{B} \neq \mathcal{A}$, then the magnitude
$\mathcal{R} = \mathcal{A} - q\mathcal{B}$ is called the remainder, and $\mathcal{B}$ exceeds $\mathcal{R}$ ($\mathcal{B} > \mathcal{R}$).

Anthyphairesis (BAFS) consists of repeated application of the division
algorithm (Proposition 2.2). Suppose that two magnitudes of the same
kind are given, a larger magnitude $\mathcal{A}$ and a smaller magnitude $\mathcal{B}$. From
Proposition 2.2 we see that there exists a natural number i such that

$$i\mathcal{B} \leq \mathcal{A} < (i+1)\mathcal{B}$$

If $i\mathcal{B} = \mathcal{A}$, stop. Otherwise, put $\mathcal{R} = \mathcal{A} - i\mathcal{B}$ and find j such that

$$j\mathcal{R} \leq \mathcal{B} < (j+1)\mathcal{R}$$

If $j\mathcal{R} = \mathcal{B}$, stop. Otherwise, put $\mathcal{S} = \mathcal{B} - j\mathcal{R}$ and find k such that

$$k\mathcal{S} \leq \mathcal{R} < (k+1)\mathcal{S}$$

and so on. This calculation may stop after finitely many steps, or it may
continue indefinitely.

The result of this process is as follows:

- A finite or infinite sequence of like magnitudes, $\mathcal{A}, \mathcal{B}, \mathcal{R}, \mathcal{S}, \dots$. The
  first two magnitudes, $\mathcal{A}$ and $\mathcal{B}$, are the given magnitudes, and $\mathcal{R}$,
  $\mathcal{S}, \dots$, are the remainders in successive applications of the division
  algorithm (Proposition 2.2).

- A finite or infinite sequence of natural numbers, i, j, k, ..., called *partial quotients*.[9]

The sequence i, j, k, ... expresses the relative size — the ratio — of the magnitudes $\mathcal{A}$ and $\mathcal{B}$. We write $\mathcal{A} : \mathcal{B} = \langle i, j, k, \ldots \rangle$. A ratio $\mathcal{A}' : \mathcal{B}'$ is equal to $\mathcal{A} : \mathcal{B}$ if the BAFS of $\mathcal{A}'$ and $\mathcal{B}'$ generates the same sequence of natural numbers $\langle i, j, k, \ldots \rangle$ as the BAFS of $\mathcal{A}$ and $\mathcal{B}$. (The Greeks would say that $\mathcal{A}' : \mathcal{B}'$ is proportional to $\mathcal{A} : \mathcal{B}$ and they would write $\mathcal{A}' : \mathcal{B}' :: \mathcal{A} : \mathcal{B}$.) The fact that $\langle i, j, k, \ldots \rangle$ *is* the ratio $\mathcal{A} : \mathcal{B}$ *arithmetizes* the concept of ratio by relating an arbitrary ratio $\mathcal{A} : \mathcal{B}$ to a sequence of natural numbers — even if the magnitudes $\mathcal{A}$ and $\mathcal{B}$ are geometric, not arithmetic, magnitudes. By this construction, we see that ratio is not a strange new beast; rather, it is a construction involving the familiar natural numbers.

The next section examines BAFS applied to numerical magnitudes.

Numerical magnitudes: the Euclidean Algorithm

The application of BAFS to a pair of natural numbers is called the *Euclidean Algorithm*. We begin with an example.

Example 2.1. Calculate the BAFS of the numerical magnitudes 871 and 403.

Solution.

$$871 = 2 \times 403 + 65 \tag{2.3}$$

$$403 = 6 \times 65 + 13 \tag{2.4}$$

$$65 = 5 \times 13 \tag{2.5}$$

Notice that it follows from equation (2.5) that 13 is a divisor of 65. Now since both terms on the right side of (2.4) are divisible by 13, it follows that 403, the left side of (2.4), is divisible by 13. Similarly, since both terms on the right side of (2.3) are divisible by 13, it follows that 871, the left side of (2.3), is divisible by 13. Thus we see that 13 is a *common divisor* of the given numbers, 871 and 403.

Furthermore, we can see as follows that 13 is the *greatest common divisor* (GCD) of 871 and 403. Equation (2.3) shows that every common divisor d of 871 and 403 must also be a divisor of 65. Moreover, (2.4) shows, since d is a divisor of 403 and 65, d must also be a divisor of 13. It follows that any common divisor of 871 and 403 must also be a divisor of 13. Since we have already seen that 13 itself is a common divisor of 871 and 403, it follows that 13 is the *greatest* common divisor of 871 and 403.

Similarly, BAFS provides a method of finding the GCD of any two natural numbers. Two natural numbers are said to be *relatively prime* if their GCD is equal to 1. The Euclidean Algorithm (BAFS of natural numbers) is introduced in Book VII of Euclid's *Elements* explicitly for the purpose of finding GCDs.[10]

Of course, one could find the GCD of 871 and 403 by finding the factorizations of both numbers: $871 = 13 \times 67$ and $403 = 13 \times 31$. This factorization technique is feasible for small numbers, but it is very difficult to factor extremely large numbers. However, BAFS (the Euclidean Algorithm) provides an easy method for computing GCDs even of very large numbers.

The BAFS of a pair of natural numbers always stops after finitely many steps. This is because the remainders (e. g., 65, 13 in the above example) become smaller at each step; a decreasing sequence of natural numbers can contain only finitely many elements. On the other hand, a BAFS of geometric magnitudes can be an infinite process—as we will see in the next section.

We can redo the calculations (2.3), (2.4), and (2.5) using fractions in a modern way not available to the Greeks, as follows:

$$\frac{871}{403} = 2 + \frac{65}{403}$$
$$\frac{403}{65} = 6 + \frac{13}{65}$$
$$\frac{65}{13} = 5$$

Putting these three equations together, we have

$$\frac{871}{403} = 2 + \frac{1}{\frac{403}{65}} = 2 + \frac{1}{6 + \frac{13}{65}} = 2 + \frac{1}{6 + \frac{1}{5}} \tag{2.6}$$

The right side of (2.6) is called a *continued fraction*. More specifically, since all the numerators are equal to 1, it is an example of a *simple* continued fraction.

Geometric magnitudes

In Book X of Euclid's *Elements*, BAFS is defined for geometric magnitudes for the purpose of distinguishing between commensurable and incommensurable magnitudes.[11] There is no direct statement in Euclid's *Elements* that the ancient Greeks also used BAFS to *define* the concept of ratio; however, some scholars[12] believe that there is ample indirect evidence to support their claim that Greek mathematicians of the fourth century BCE, Theaetetus and others, used BAFS to define ratio.

To simplify the discussion of geometric magnitudes, we consider only linear magnitudes. By *linear magnitude* we mean the total length of a figure consisting entirely of line segments. The Greek theory of ratio and proportion for this kind of magnitude is essentially equivalent to the modern theory of real numbers.[13] Admitting other types of magnitudes does not carry us further.

We begin with an example that returns to the question of incommensurability of the diagonal and side of a square (Proposition 2.1).

Example 2.2. Find the ratio of the diagonal and side of a square.

The ratio is defined by the sequence of partial quotients generated by the BAFS procedure.

We will see two solutions, one ancient and one modern. The first solution is from Fowler (1987), where it is described as an interpretation of material from the commentary of Proclus (411–485 CE) on Plato's *Republic*. This solution uses methods that were available to Greek mathematicians of the fourth century BCE.

Ancient Solution. Figure 2.4 is constructed starting from the diagonally placed shaded square AEFG. The side of the larger square ABCD is equal to the side plus the diagonal of the smaller square so that $\overline{BF}$ is equal to $\overline{EF}$. From Figure 2.4 it is clear that the diagonal AC of the large square is equal to the diagonal plus twice the side of the small square so that $\overline{CG} = \overline{AD}$. The first step in the BAFS of AC and AD is

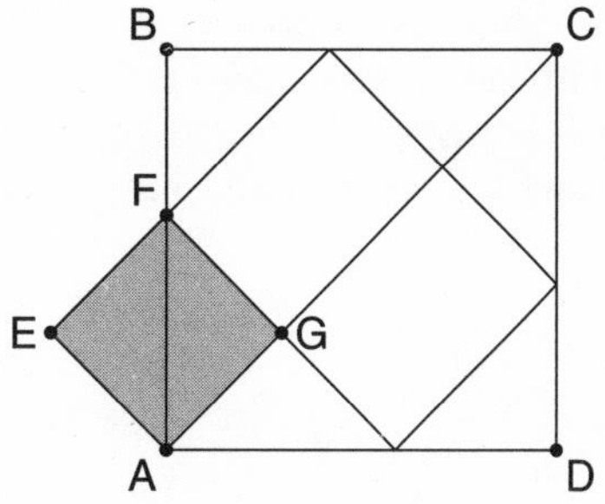

Figure 2.4.

$$\overline{AC} = 1 \times \overline{CG} + \overline{AD}$$

The first partial quotient is equal to 1.

For the second step of the BAFS, we must apply the division algorithm to AD and AG. Now we use the fact that the side of the large square is equal to the diagonal plus the side of the small square:

$$\overline{AD} = \overline{AF} + \overline{AG}$$

Since we are only interested in the next partial quotient, the size of the square is not important. For an *arbitrary* square—for example, the large square ABCD—the division algorithm of the side plus the diagonal $(\overline{AD} + \overline{AC})$ with respect to the side the square $(\overline{AD})$ yields a result that is independent of the size of the square. More precisely, applying the division algorithm to $\overline{AD} + \overline{AC}$ and $\overline{AC}$, we obtain

$$\overline{AD} + \overline{AC} = 2 \times \overline{AD} + \overline{AG}$$

Hence, the second partial quotient is 2.

For the third step of the BAFS, we perform the division algorithm for AD with respect to AG. But this is exactly the same calculation that we made above in step 2. Therefore, the third partial quotient is 2. Continuing in the same fashion, all of the succeeding partial quotients are also equal to 2. Thus the sequence of partial quotients for the ratio of the diagonal to the side of square is $\langle 1, 2, 2, 2, \ldots \rangle$.

An alternate method of proof is to relate the steps of the BAFS to ever smaller squares, shown in Figure 2.5.

Modern Solution.   We want to compute BAFS of $\sqrt{2}$ with respect to 1. We begin with a numerical experiment, shown in Table 2.1, using the decimal approximation $\sqrt{2} \approx 1.414214$. The resulting arithmetic calculations are not difficult for us, but they would have been impossible for the ancient Greeks. Although this experiment may put us on the right track, it fails to be a self-contained rigorous demonstration because we have not derived this approximation of $\sqrt{2}$, and, after all, it is only an approximation.

This calculation suggests that the partial quotients for the BAFS of $\sqrt{2}$ with respect to 1 are $\langle 1, 2, 2, 2, \ldots \rangle$. We can prove this rigorously with the following algebraic calculation—the same calculation as above but using exact values instead of approximations. We make repeated use of the following identity:

$$\sqrt{2} - 1 = \frac{(\sqrt{2} - 1)(\sqrt{2} + 1)}{\sqrt{2} + 1} = \frac{1}{\sqrt{2} + 1}$$

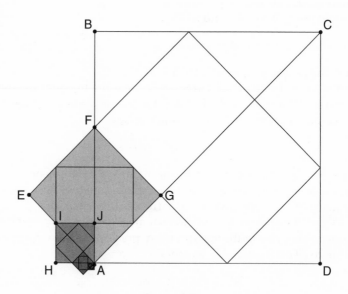

Figure 2.5.

Table 2.1. Numerical approximation of BAFS of $\sqrt{2}$ with respect to 1.

$1.414214 - \mathbf{1} \times 1.000000 = 0.414214$
$1.000000 - \mathbf{2} \times 0.414214 = 0.171572$
$0.414214 - \mathbf{2} \times 0.171572 = 0.071070$
$0.171572 - \mathbf{2} \times 0.071070 = 0.029432$
$0.071070 - \mathbf{2} \times 0.029432 = 0.012206$
$0.029432 - \mathbf{2} \times 0.012206 = 0.005020$
$0.012206 - \mathbf{2} \times 0.005020 = 0.002166$
$\cdot \qquad \cdot \qquad \cdot$

Here is the exact computation of the BAFS of $\sqrt{2}$ and 1:

$$\sqrt{2} - \mathbf{1} \times 1 = \frac{1}{\sqrt{2}+1}$$

$$1 - \mathbf{2} \times \frac{1}{\sqrt{2}+1} = \frac{(\sqrt{2}+1)-2}{\sqrt{2}+1} = \frac{\sqrt{2}-1}{\sqrt{2}+1} = \frac{1}{(\sqrt{2}+1)^2}$$

$$\frac{1}{\sqrt{2}+1} - \mathbf{2} \times \frac{1}{(\sqrt{2}+1)^2} = \frac{(\sqrt{2}+1)-2}{(\sqrt{2}+1)^2} = \frac{\sqrt{2}-1}{(\sqrt{2}+1)^2} = \frac{1}{(\sqrt{2}+1)^3}$$

$$\frac{1}{(\sqrt{2}+1)^2} - \mathbf{2} \times \frac{1}{(\sqrt{2}+1)^3} = \frac{1}{(\sqrt{2}+1)^4}$$

$$\frac{1}{(\sqrt{2}+1)^3} - \mathbf{2} \times \frac{1}{(\sqrt{2}+1)^4} = \frac{1}{(\sqrt{2}+1)^5} \cdots$$

This calculation confirms the correctness of the partial quotients suggested by our numerical experiment. It also implies that the continued fraction expansion of $\sqrt{2}$ is as follows:[14]

$$1 + \cfrac{1}{2 + \cfrac{1}{2 + \cfrac{1}{2 + \cdots}}}$$

In this chapter, we have seen how questions of incommensurability led ancient Greek mathematicians to develop ideas of magnitude, ratio, and proportion. We have seen two Greek definitions of ratio—important extensions of the number system.

The rational numbers don't have visible gaps; between any two there are infinitely many more. Yet, there is a gap where $\sqrt{2}$ should be, and this was puzzling to the Greeks. The Greek theories of ratio and proportion found a way to bridge this gap, making an important advance toward the modern real number system. But in roughly two thousand years between then and now, much remained to be done.

Of course, negative numbers and the decimal system needed to be invented, together with long division and other algorithms. More fundamentally, the Greeks failed to see that it is not necessary to have an assortment of different magnitude concepts — one for lengths, another for areas, and so on. Today, the needs of science are served by a single concept of magnitude: the real numbers.

The next chapter discusses musical applications of the theory of ratio and proportion. We will see how today's science and musical practice has extended and refined the ancient Greek belief that certain ratios define pleasing musical intervals.

3

The Music of the Ratios

*There is geometry in the humming of the strings; there is music in the
spacing of the spheres.*

—PYTHAGORAS, 582?–500? BCE

IN THE SIXTH CENTURY BCE, PYTHAGORAS and his followers believed
that numbers are the language in which the meaning of the universe
is written. Numbers are also fundamental in today's world in which
digital computers—containing streams of 0s and 1s—entertain us, perform our accounting needs, and compute the interaction of galaxies. Although the Pythagoreans lacked the digital computer, their belief in the
universal importance of number was confirmed by devices familiar to
them—musical instruments. In fact, Aristotle tells us that the Pythagoreans "supposed the elements of numbers to be the elements of all things,
and the whole heaven to be a musical scale and a number."[1]

In the hands of the Pythagoreans, a simple musical instrument became a scientific instrument. We infer this from their teachings, although
we have no ancient log books detailing their actual experiments. Unfortunately, after this promising beginning, experimental science came to a
standstill in ancient Greece. This was due in part to the influence of philoso-
phers like Plato, who were more impressed by the unsupported speculations of the Pythagoreans than by their empirical observations. Plato
taught that the universe is perfect; and, therefore, empirical investigation
is unnecessary because to discover the nature of the universe it is sufficient to examine, through philosophical discourse, the general concept of
perfection. Circles, spheres, and ratios of small whole numbers were considered obvious instances of perfection, and therefore, it was argued, the
motion of the heavenly bodies—the sun, moon, planets, and stars—must
be based on these concepts. Two thousand years later, in the late Renaissance, empirical science—the basis of all modern science—was finally

rediscovered by Galileo and others. Galileo not only revised the Pythagoreans' theories of the heavens; he also had a surprising familial connection with their musical ideas. More specifically, his father, Vincenzo Galilei (1520?–91), was a noted lutenist and composer who joined the controversy concerning the merits of Pythagorean tuning of his instrument. In his dialog, *Fronimo* (1584), Vincenzo favored *equal-tempered* tuning, which we will discuss later.

We can duplicate the Pythagoreans' observations using any pair of stringed instruments—a pair of violins or guitars will do. We wish to bow or pluck two strings of different lengths simultaneously, but we require that the two strings are made of exactly the same material and are under precisely the same tension. Figure 3.1 shows a hypothetical instrument that would be especially suitable for the Pythagorean experiments.

Although no one knows how ancient Greek music actually sounded, we do know that the Pythagoreans made a connection between music and mathematics. The Pythagoreans discovered that harmonious musical intervals are generated by pairs of string lengths with simple numerical ratios. These intervals are called *Pythagorean* to distinguish them from the slightly different *equal-tempered* intervals in current musical use. In particular, halving the string length raises the pitch by exactly one octave, and if the ratio of the string lengths is 2 : 3, then the corresponding pitches form an interval called a *Pythagorean* or *perfect fifth*—so named because the interval *do–sol* between the first and *fifth* notes of the diatonic[2] scale is an instance of this interval. Other important Pythagorean intervals correspond to the ratios 3 : 4 (the *fourth*), 8 : 9 (the *major second*), and 9 : 16 (the *minor seventh*).

The role of numerical ratios in music might seem mildly interesting to us, but to the Pythagoreans it was astonishing—even staggering—and emboldened them to speculate that the same numerical ratios controlled the motion of the heavenly bodies. Their musical experiments led the Pythagoreans to believe that the small whole numbers were the key to understanding the universe and that music emanated from the heavenly bodies. Only Pythagoras himself, because of his higher level of spirituality, was presumed to have actually heard the music of the spheres.

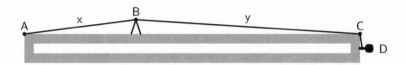

Figure 3.1. A musical instrument for the Pythagorean experiments. A single string AC is stretched across the bridge B, which can be moved to adjust the ratio x : y. Tension on the string can be adjusted using the tuning peg at D. The two parts of the string with lengths x and y are plucked simultaneously.

While the Pythagoreans overestimated the importance of their discoveries, they raised questions that have been answered only recently. The Pythagorean experiments initiated the science of acoustics—more specifically *psychoacoustics*, the study of the *perception* of sound. If a tree falls in the forest and no one hears it, the sound is strictly in the domain of acoustics. But if someone *does* hear it, it enters the purview of psychoacoustics. It is beyond the scope of this book to discuss these topics exhaustively. We confine our attention to matters that relate to the musical observations of the Pythagoreans. In particular, to discuss consonance and dissonance, it is sufficient to consider only sounds that consist of one or more *sustained tones*—sounds that do not change over time. We do not consider transient sounds or noise.

Were the Pythagoreans right? Is it true that certain sounds are intrinsically more musical than others? Or are our musical preferences purely a product of our culture? We hope to answer these questions.

The Pythagorean ideas of harmony lead us to the physics of sound and the anatomy of the ear. We will see a mathematical foundation for the familiar music of today and the forgotten music of the Pythagoreans.

Acoustics

Acoustics is the physics of sound. Physically, sound is a rapidly varying pressure wave propagated through air, water, or other media. The rate of vibration of audible sound is between 20 and 20,000 cycles per second. In air at 68°F, sound waves travel at a speed of 770 miles per hour.

An ear or a microphone detects the changing pressure at a particular point in space. The ear transmits this information to the brain. The electrical output from a microphone can be used to activate a loudspeaker, but it might also generate a graph of pressure versus time.

The simplest musical sound is called a *simple tone* or a *sine tone*. The pressure–time graph generated by a simple tone is a *sine wave*, as shown in Figure 3.2. The horizontal coordinate is time, and the vertical coordinate is pressure, which varies about an equilibrium steady-state value.

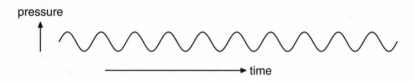

Figure 3.2. The pressure–time graph of a simple tone is a *sine wave*.

The rotating circle

A sine wave is more than just any wiggly curve. It is generated by certain precise mathematical relationships. We can make a model of a sine curve by cutting a cylindrical roll of paper by a diagonal plane. By unrolling the paper flat, we obtain a sine curve. The vibration of a mass suspended by an ideal spring[3] is described by a sine curve. Figure 3.3 shows an interpretation of a sine curve using a rotating circle. This construction shows that a sine wave is specified by three quantities—amplitude, period, and phase:

1. The *amplitude* is the maximum vertical deviation from the steady state—half the vertical distance between a trough and a peak. The larger the amplitude, the louder the sound.

The mean pressure of the loudest sound that the ear can tolerate is about 2,000,000 times greater than the mean pressure of a barely perceptible sound. This enormous range makes pressure an unsuitable unit of measurement of loudness. The standard unit of loudness is the decibel (dB).[4] Increasing loudness by 1 decibel is approximately equal to a 26% increase in mean pressure. An increase of 10 decibels is equivalent to multiplying mean pressure by 10 exactly. Logarithmic units, like the decibel, are useful when the quantity measured encompasses many orders of magnitude. Decibels are a better measure of *perceived* loudness than pressure.

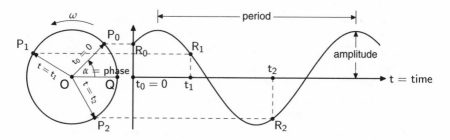

Figure 3.3. A rotating circle generates a sine curve. The circle with center at O rotates counterclockwise at constant angular velocity ω. The sine curve is a graph of the time dependence of the vertical height of a point fixed to the rotating circle—a point that is initially at P_0. The position of this point at times t_1 and t_2 are P_1 and P_2, respectively. The dashed lines show how these points on the circle determine points on the sine curve.

The *amplitude* is equal to the maximum height of the sine curve—the radius of the circle. The *period* is the distance between successive peaks of the sine curve—the time for one revolution of the circle. The *frequency* (in *Hz*) is the angular velocity of the moving point measured in revolutions per second. The *phase* is the angle α between OP_0 and the horizontal radius OQ.

2. The *period* is the horizontal distance (time) between two adjacent peaks (or troughs).

The reciprocal of the period is the *frequency*—the number of peaks (or troughs) in unit time. Hence, the frequency of a sine curve contains the same information as the period. Frequency is measured in *cycles per second*—also called *hertz*—after Heinrich Rudolf Hertz (1857–94), the German physicist who was the first to transmit and receive radio waves. The higher the frequency, the higher the pitch.

There is an analogy between mean pressure (loudness) and frequency. In both cases the ear perceives percent change rather than absolute change. Musical intervals are defined by frequency *ratios*. For example, an upward semitone is about a 6% increase in frequency. In music, logarithmic units of frequency are more suitable than hertz. All of the standard musical intervals—fourths, fifths, octaves, and so on—are logarithmic units of frequency.

3. The *phase* of a sine wave, as shown in Figure 3.3, defines the starting position of the sine wave. If two sine waves have the same amplitude and frequency, it is always possible to bring the one into coincidence with the other by means of a horizonal shift—a phase shift. The amount of this shift is called the phase difference.

The Pythagoreans did not have the technology to discover that the pitch of a sound corresponds to the frequency of a vibration. Nevertheless, the Pythagorean ratios apply to frequency as well as the length of a string. For example, doubling the length of a string halves the frequency of vibration. However, the sounds heard from a vibrating string are not simple tones, but rather a mixture of simultaneous simple tones. This happens because a string has not just one but a sequence of modes of vibration. Sometimes the higher modes are used for musical effect—for example, when a violinist places a finger lightly on a node of the vibrating part of a string. The first three modes of vibration of a string are shown in Figure 3.4.

In the first mode (a)—also called the *fundamental* mode—the string entire alternately bows upward and downward. The second mode (b) is one octave above the fundamental; the third mode (c) is an octave plus a perfect fifth above the fundamental; the fourth mode is two octaves above the fundamental; the fifth mode is two octaves plus a Pythagorean major third above the fundamental; and so on. The fundamental together with

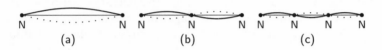

Figure 3.4. The first three modes of vibration of a stretched string. The stationary points labeled N are called nodes.

the higher modes of vibration are called *partials*. The partials higher than the fundamental are called *overtones*. The mixture of amplitudes of the various partials gives a quality called *timbre* to the tone. It is the timbre that tells us that we are hearing a violin and not a guitar.

Integer multiples, two or greater, of a given frequency are called *harmonics* of that frequency. The string, woodwind, and brass instruments have harmonic partials. However, the drum, the xylophone, the bells, and most other percussion instruments have *nonharmonic* partials that are not integer multiples of a fundamental frequency. We will see in the discussion on psychoacoustics that the consonance of the Pythagorean intervals stems from the fact that the higher partial frequencies of a vibrating string are all harmonics of the fundamental.

Woodwind instruments also have harmonic partials. The partials of those woodwind instruments that are effectively pipes open at both ends, for example, the flute and the recorder, tend to include both even and odd harmonics. Despite appearances, the oboe and the saxophone are also effectively open pipes. However, some woodwinds that function as pipes closed at one end (e.g., the clarinet) suppress the *even* harmonics. This accounts for the unique timbre of the clarinet. For woodwind instruments, the fingerings from the lowest note up to its first overtone form a cycle that is, with small modifications, repeated for higher notes. For all woodwinds, a variety of tones are formed by opening and closing holes in the side of the tube—to change the effective length of the pipe and to suppress fundamental tones in favor of the higher partials. For the flute, the first overtone is an octave above the fundamental, and the cycle consists of 12 fingerings for the 12 semitones. The fingerings for the second octave are identical or similar to those of the first octave; the flutist can play either the fundamental tone or the first overtone, the octave, by adjusting the air pressure and the tension of the lips. For the clarinet, on the other hand, the first strong overtone is an octave *plus a fifth* above the fundamental, and, therefore, there is a cycle of 19 different fingerings corresponding to the 19 distinct semitones between the lowest note and its first overtone. Thus, the clarinetist needs to learn a longer cycle of fingerings than the flutist. On the other hand, for the clarinet, two cycles includes more than three octaves, whereas to achieve the third octave the flutist needs to learn additional special fingerings.

The drum is an example of a musical instrument that has nonharmonic partials. A drum produces sound by means of a vibrating membrane. The fundamental tone for a drum corresponds to a back-and-forth vibration of the entire drum surface. For a circular drum, the lowest pitched partial above the fundamental corresponds to a vibration in which a diameter is a nodal line that bounds two semicircular regions that vibrate in opposite phase. The frequency of this partial is 1.594 times the frequency of the fundamental. For example, if the fundamental is C, then this lowest partial is slightly higher than G-sharp above C, a minor sixth. The corresponding lowest overtone for the vibrating string is exactly an octave. The partials of a drum are far more complex. The frequency factors for a few of the partials of a circular drum are listed below in increasing order:

1.000, 1.594, 2.136, 2.296, 2.653, 2.918, 3.156, 3.501, 3.600, 3.652, 4.060, · · ·

Each number in this list corresponds to a different pattern of nodal lines on the circular drum. (For the vibrating string, the corresponding numbers are 1, 2, 3,....)

In 1956, mathematician Mark Kac asked, "Can you hear the shape of a drum?" This is a clever way of asking if the sequence of partials of a drum implies exactly one shape for the drum's membrane. In particular, do the frequency factors listed above somehow imply that they are generated by a *circular* drum? The answer to Kac's question is negative, because in 1991 mathematicians Carolyn S. Gordon and David L. Webb showed that there exists a pair of differently shaped drums that generate the same sequence of partials.

In addition to the drum, most percussion instruments produce non-harmonic partials: the xylophone, the bells, and the cymbal. Of course, *nonharmonic* does not mean dissonant or unpleasant.

Waveforms and spectra

Periodic waveforms

A sine wave (Figure 3.3) is an example of a *periodic waveform*. It is called periodic because it is repetitive over time. It is sufficient to depict the graph over a time interval equal to one period. Shifting the initial period horizontally, we can obtain as much of the curve as needed.

Using electronic devices, we can generate and hear a great variety of periodic waveforms. For example, an electronic organ can simulate a variety of different musical instruments. With a soldering iron and parts readily available from an electronics store, a hobbyist can generate waveforms in his or her garage. The simplest project of this sort is the construction of a *multivibrator*, a device that produces a *square wave*, the waveform shown in Figure 3.5. A computer requires millions of transistors, but a multivibrator just two. Plans are available in books, magazines, and websites for electronics hobbyists. Electronic waveforms can be viewed with an *oscilloscope*, a device that uses a cathode ray tube — now used as the picture tube in television sets. The oscilloscope was invented by Karl Ferdinand Braun (1850–1918) in 1897, 50 years before the advent of television.

It is very remarkable that a square wave — in fact, an arbitrary periodic waveform — can be approximated as close as needed by a series of sine waves.[5] The series consists of a sine wave with the same period as the

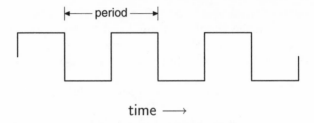

Figure 3.5. A square wave.

square wave together with harmonics of this fundamental frequency with
suitably chosen amplitudes and phase angles. The square wave is approximated by *sums* of these wave forms.[6] For an acoustical square wave, the
approximating sine waves are *partials*.

1. *Frequency.* For a square wave, the frequencies of the partials are
   odd multiples only of the fundamental. Recall that the clarinet also
   has principally odd harmonic partials. There are websites where the
   sound of a square wave can be heard. The sound of a square wave is
   vaguely like a clarinet, but far from pleasant—the ugly cousin of the
   clarinet. On the other hand, a square wave generator can do something that a clarinet cannot do—trouble-shoot a radio or TV.
2. *Amplitude.* For the square wave, the amplitudes of the odd harmonics of order 1, 3, 5,... are proportional to the reciprocals

$$1, \, 1/3, \, 1/5, \ldots$$

   respectively. This sequence of reciprocals drops off much more slowly
   than the corresponding sequence for a musical instrument. This is
   why we say that the square wave is "rich" in harmonics.
3. *Phase.* For the square wave, the phase angle of each partial is $0°$.
   (See Figure 3.6(a).) Altering the phases of the partials generates a
   different waveform. For example, Figure 3.6(b) shows the waveform
   generated by the sum of these 6 partials if the phase angles are set to
   $90°$ instead of $0°$. To the ear, the waveforms of Figures 3.6(a) and (b)
   sound exactly the same because they have the same partial frequencies with the same amplitudes. The waveforms, the solid curves, in
   Figures 3.6(a) and (b) are different, yet we cannot *hear* the difference.
   This is an instance of *Ohm's law of hearing*, after the German physicist Georg Simon Ohm (1789–1854),[7] which states that the ear generally cannot detect alterations of phase in the pure components of a
   complex sound. On the other hand, changes in the frequencies and
   amplitudes of the pure components of a complex sound are readily
   perceived by the ear.

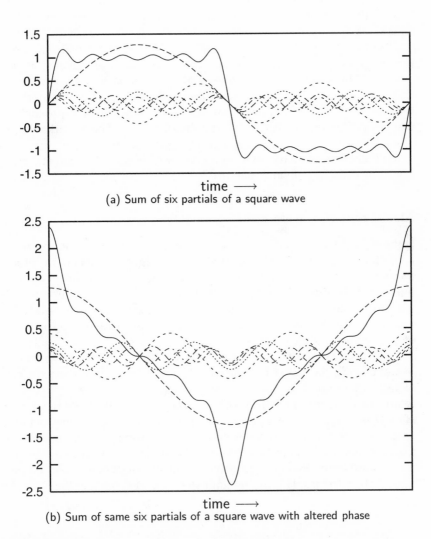

(a) Sum of six partials of a square wave

(b) Sum of same six partials of a square wave with altered phase

Figure 3.6. Approximating a square wave by adding six of its partials. In (a), the six partials are indicated by the dotted curves. The solid curve, the sum of the six partials, resembles the square wave. The phase for each of these partials is 0°. In (b), the dotted curves are the same partials as in (a) but with phase angles changed from 0° to 90°; and the solid curve is the sum of these partials. In accordance with Ohm's law of hearing, the ear cannot perceive any difference between the sounds generated by the waveforms (a) and (b).

In Figure 3.6(a), the dotted sine waves represent the first six partials of a square wave, and the solid curve represents the sum of these six sine waves. From these graphs, it is plausible that this sum approximates the square wave.

The foregoing illustrates a general fact. A periodic waveform subject to very general regularity conditions can always be approximated as close as desired by a series of sine waves that are harmonics of the fundamental frequency with suitably chosen amplitudes and phases. Furthermore, there is only one such approximating series. This series approximating waveforms by sine waves is called a *Fourier series*. The remarkable idea that arbitrary periodic waveforms can be approximated in this way is due to the French mathematician Joseph Fourier (1768–1818), friend and scientific advisor to Napoleon Bonaparte.[8]

Almost periodic waveforms

Is the sum of two periodic waveforms periodic? No, not if the two waveforms have incommensurable periods — as in Figure 3.7, where the ratio of the periods is $1 : \pi$. This figure does not exhibit any obvious periodicity, but, of course, this does not prove that the waveform is not periodic. The waveform shown here can be generated by two simple tones played simultaneously. We need to include this sort of waveform in our discussion of consonance and dissonance.

The class of periodic waveforms is too confining. We need a wider class of waveforms, the *almost periodic* waveforms. For our purposes, the class of almost periodic waveforms consists of those that can be approximated to arbitrary precision by sums of sine waves — with possibly incommensurable periods. Unlike a periodic waveform, we cannot construct all of an almost periodic waveform by endlessly repeating part of its graph. The theory of almost periodic functions was created by the Danish mathematician Harald Bohr (1887–1951), brother of the physicist Niels Bohr. It is said that he always started his mathematics lectures by writing at the upper left

time $\longrightarrow$

Figure 3.7. A nonperiodic waveform, a sum of two sine waves with periods in the ratio $1 : \pi$.

corner of the blackboard and finished at the lower right corner exactly 50 minutes later.

Almost periodic waveforms are represented by a unique series of sine waves much the same as periodic waveforms. In general, almost periodic waveforms have nonharmonic partials. That is, the frequencies need not be integer multiples of a fundamental frequency. The waveforms generated by percussion instruments (e.g., drum, xylophone, bells, chimes) are almost periodic but not periodic. Bohr's theory of almost periodic functions is the proper mathematical foundation for the acoustics of musical tones.

Spectra

The fact that the sounds represented by the waveforms in Figure 3.6(a) and (b) cannot be distinguished by ear demonstrates that waveforms are not very useful in the study of hearing. We need a method of presenting information about a tone that identifies frequencies and amplitudes and ignores phase. The spectrum of a tone shows precisely this information. Figure 3.8 compares the spectrum of C4 (middle C — 262 Hz) as rendered

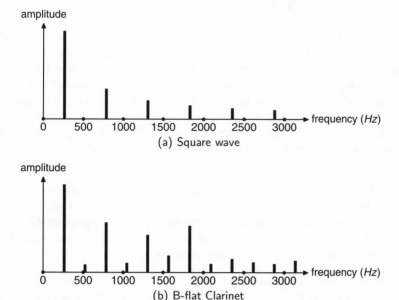

Figure 3.8. Spectrum of a fundamental tone of 262 Hz (middle C) rendered by (a) a square wave generator and (b) a B-flat clarinet — showing partials up to 3000 Hz. The scale of amplitude is linear (not decibels). No amplitude scale is shown because only relative amplitude is important.

by (a) a square wave generator and (b) a B-flat clarinet. The musical term for spectrum is *timbre*. The ear can detect the distinctive timbre of the clarinet or the flute.

A sound spectrum breaks down a complex tone into the frequency components, together with the corresponding amplitudes. A device that achieves this decomposition is called a *frequency analyzer*. The *Helmholtz resonator*, named after the German scientist Hermann von Helmholtz (1821–94),[9] is the oldest and simplest frequency analyzer. Helmholtz found that a spherical vessel with suitable dimensions resonates at a particular frequency with very little contamination from nearby frequencies. In shape, a Helmholtz resonator resembles the spherical Florence flask used in chemical laboratories. A short tube connected to the resonator can be fitted snugly into the ear using soft wax. Each resonator enables one to hear whether a particular frequency is a partial of a complex tone. A separate resonator is required to monitor each frequency of interest.

The Helmholtz resonator is primitive by the standard of current technology. Now frequency analyzers convert sound to an electric vibration that is decomposed into its sinusoidal components using electronic filters.

Psychoacoustics

How do we hear pitch? Helmholtz opened the door to the modern chapter of psychoacoustics by declaring that the ear is a frequency analyzer. In doing so, he opposed many of his contemporaries — including his mentor, the physiologist Johannes Müller — who held to the philosophical principle of *vitalism*, which says that living processes can never be completely understood through the application of ordinary physics, chemistry, and mathematics. Although Helmholtz did not discover the precise mechanism of the ear's frequency analyzer, his assertion was much more than speculation:

- It is clear that the ear does perform frequency discrimination. Indeed, a musician with perfect pitch can identify middle C. (Yet, no one seems able to identify all the partials of a complex tone.)
- It seems reasonable that the ear makes sense of complex sounds by reducing them to simpler components.

In 1961, the Hungarian-born American researcher Georg von Békésy (1899–1972) won the Nobel prize in medicine for showing how the ear functions as a frequency analyzer. The inner ear contains a snail-shaped spiral structure the size of the tip of a little finger, the *cochlea*. The cochlea contains a coiled structure about 1.3 inches long called the *basilar membrane*. Because of its varying width and stiffness, different frequencies cause different parts of the basilar membrane to bulge. The bulging of the basilar membrane activates nerves contained in the *organ of Corti*, sending information to the hearing center of the brain. The foregoing explanation is called the *place* theory of pitch perception because it explains how different simple tones are associated with different sites on the basilar membrane. Despite some difficulties, the place theory is considered the primary explanation of pitch perception.

Although the ear is a wonderfully sophisticated instrument, it shares a certain limitation with all frequency analyzers—a limit of resolution. If the frequencies of two simultaneous simple tones are sufficiently far apart, they are heard as separate tones, but they are perceived in a more complex way, to be discussed later, if they are close enough to be in the same *critical bandwidth*.

The critical bandwidth underlies the perception of consonance and dissonance. The interplay of consonance and dissonance is one source of music's endless fascination.

Consonance versus dissonance

We hear tinkling, pounding, buzzing, whirring, roaring, and, most important of all, the sound of speech. However, our concern here is mainly sound as it is perceived by musicians and their listeners. More particularly, we consider how we hear two simultaneous sustained musical tones.

Our ears recognize that the pair of tones C–G has the same character as D–A; we say both are fifths. The Pythagoreans discovered that the fifth, the octave, and other harmonious intervals, are determined by ratios of string lengths, for example, 3 : 2 and 2 : 1. In modern terms, each interval derives its identity from the ratio of the frequencies of the two tones that generate it. It is the ratio of the frequencies that somehow tells us that we are hearing a fifth, an octave, or some other interval.

By listening to simultaneous tones on stringed instruments, the Pythagoreans found certain musical intervals harmonious and others dissonant. In this section, we will see that if they had studied drums instead of stringed instruments, they might not have discovered consonance based on simple ratios.

The foundational theory of consonance and dissonance was developed by Helmholtz. The next section considers results of Plomp and Levelt (1965) that confirm and amplify Helmholtz's work. One might start with the simplest instance of a harmonious interval—the octave—and argue that any explanation of consonance must confirm that an octave is always a consonant interval. It turns out that this plausible idea is a bad start. The partials of the tones of musical instruments confuse the matter. It turns out that the octave is especially consonant if all the partials are harmonics of the fundamental—a condition that fails for the drum and other percussion instruments.

Critical bandwidth

Instead of considering particular intervals, such as the octave, it is better to start by examining the question of consonance–dissonance for *simple tones* with no higher partials. Simple tones can be generated by an electronic audio oscillator. There are three principal ways in which we hear two simultaneous simple tones:

1. If the pitches of the two tones are far apart, then we hear two tones sounded simultaneously, and the addition of the second tone increases our sense of loudness. Two tones are heard if the frequencies of the two tones are farther apart than the *critical bandwidth* — the smallest frequency difference at which two tones are judged completely consonant. (See Figure 3.9.)

2. If the pitches of the two tones are sufficiently close together, we hear the two tones fused into an intermediate pitch that varies in loudness in a periodic fashion. This phenomenon is known as *beats*. The frequency

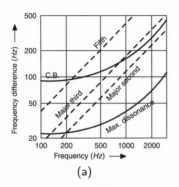
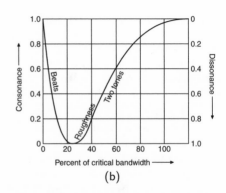

(a) (b)

Figure 3.9. (a) The *critical bandwidth* (C.B.) is the smallest frequency difference such that two simultaneous simple tones are heard as consonant separate tones. If two simple tones are closer but not identical, they are heard as beats, roughness, or two more or less dissonant tones. The critical bandwidth depends on frequency. Using logarithmic horizontal and vertical scales, this graph shows, for each given frequency, the widths of several frequency intervals with the given frequency as midpoint: (1) the critical bandwidth; (2) the width of the interval of maximum dissonance, 25% of the critical bandwidth. (3) The width of the intervals of the *fifth, major third,* and *major second.* Note that these standard musical intervals are perceived more dissonant at lower frequencies.

(b) The *perception of dissonance* varies as the frequency difference between two simultaneous simple tones is varied from 0 up to the critical bandwidth. This graph shows the perception of consonance–dissonance on a scale from 0 to 1. For example, at about 25% of the critical band width, the sound is perceived as maximally dissonant.

Based on Plomp and Levelt (1965, Figures 8, 9, and 10). With permission, *Acoustical Society of America,* Copyright 1965.

of the beats is equal to the difference in the frequencies of the two tones. The addition of the second tone does not increase our sense of loudness. Beats occur because the two tones alternately reinforce and cancel each other — that is, they are alternately in and out of phase. Figure 3.10 shows the pressure–time graph of an instance of beats.

3. Two tones can be too close together to be heard as separate tones and too far apart to be heard as beats. This type of sound is considered most dissonant and is generally called *rough*.

Plomp and Levelt asked subjects[10] to evaluate the consonance and dissonance of pairs of simple tones sounded simultaneously. As the frequency gap between the tones gradually widened, subjects initially heard a single tone with beats and reported a gradual increase in dissonance. Dissonance reached a peak as beats gave way to roughness. As the gap was widened further, the sensation of roughness diminished and two separate tones were gradually heard; consonance gradually increased to the maximum value. Further increases in the frequency gap continued to be evaluated as consonant. Remarkably, the subjects did not find the traditional musical intervals any more consonant than other nearby intervals. For example, subjects did not consider a true octave more consonant than an "out-of-tune" octave.

These results seem to contradict the findings of the Pythagoreans. However, the Pythagoreans made their observations using stringed instruments, and, as shown above, a plucked string does not generate a simple tone because it has *harmonic partials*. In fact, Plomp and Levelt *confirmed* the Pythagorean observations by administering further tests using tones with harmonic partials. On hearing such tones, subjects reported that, for example, Pythagorean fourths, fifths, and octaves were more consonant than

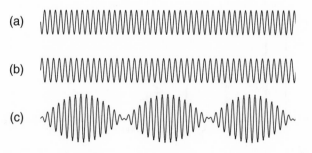

Figure 3.10. Beats. Graphs (a) and (b) show pressure versus time for a pair of simple tones with frequencies 310 Hz and 290 Hz, respectively, over a time period of 0.2 sec. (It is barely perceptible to visual inspection that (a) has more peaks and troughs than (b).) When the simple tones are sounded simultaneously, the pressures are alternately in and out of phase, producing *beats* (c), which pulsate at a rate equal the frequency difference of the two tones, in this case 20 Hz.

out-of-tune versions of these intervals. It is surprising that changing the test tones from simple to complex should make such a radical difference. This striking change occurs because a single tone with many partials is equivalent to an entire chorus of simple tones. When two complex tones are sounded together, a dissonance may result from the interaction of any pair of partials when both are within a single critical bandwidth.

Even a single tone with harmonic partials can be dissonant. Very high harmonics can be sufficiently close in frequency to create dissonance. For example, for a tone of 100 Hz, the fundamental and first harmonic (200 Hz) is an octave, but the interval between 16th and 17th harmonics (1700–1800 Hz) is close to a minor second, a dissonant interval. A single tone created by an electronic square wave generator sounds unpleasant because the high harmonics have high amplitude—a characteristic that makes the multivibrator a useful electronic test instrument.

Figure 3.11 shows that middle C, played as a single tone on a clarinet, has dissonant components. This figure is an elaboration of Figure 3.8(b), with the critical bandwidth associated with each partial shown as a gray extension of the spectral lines.

Harmonic partials of two simultaneous tones at an interval larger than the critical bandwidth can create dissonance. For example, a major seventh is dissonant if the lower tone has a harmonic partial one octave above the fundamental because this partial is too close to the fundamental of the higher tone. An octave is consonant if both tones have only harmonic partials because all of the partials of the upper tone coincide with partials of the lower tone.

The dependence of the critical bandwidth on the mean frequency of the two tones is shown in Figure 3.9(a). As the frequency of the two tones increases, the critical bandwidth, measured in hertz, increases. Neverthe-

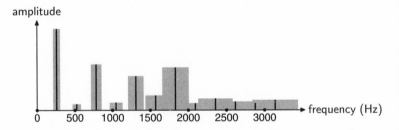

Figure 3.11. Spectrum of a fundamental tone of 262 Hz (middle C) rendered by a B-flat clarinet. (See also Figure 3.8(b).) The gray extension of the spectral lines indicates the critical bandwidth associated with each partial frequency. The critical bandwidths of the higher partials overlap. This shows that the timbre of the clarinet has dissonant elements. The partials that contribute to this dissonance have relatively low amplitudes. Most musical instruments exhibit similar low-level dissonance.

less, critical bandwidth measured in *semitones* decreases as frequency increases. This is why, in Figure 3.9(a), certain standard musical intervals become more dissonant at lower frequencies. Despite this fact, musicians sometimes mistakenly think that the consonance or dissonance of an interval is determined purely by the number of semitones it contains—that thirds are consonant and seconds dissonant in every octave, high or low. On the contrary, most composers demonstrate at least an intuitive understanding that this is not true. Their compositions tend to avoid close intervals in the bass because their instinct, based on musical experience, tells them that such intervals tend to be dissonant.

Intervals, Scales, and Tuning

In the Western musical tradition, the octave is the musical North Star because for the dominant Western musical instruments, the strings, the winds, and the singing voice, the octave is the most consonant of intervals. By contrast, in the Indonesian musical tradition, which is dominated by the nonharmonic timbre of percussion instruments, the octave is less important. In this section, we examine intervals, scales, and tuning in Western music.

For Western tuning systems, the octave is fixed, and unchanging. A *scale* is a method of interpolating pitches within the octave.

The perfect fifth is next to the octave in importance. The Pythagorean and equal-tempered tuning systems are based on the fact that the interval of 12 perfect fifths is *almost* the same as seven octaves. We will see that this is the reason why the chromatic scale has 12 tones.

Pythagorean tuning

The Pythagoreans did not have keyboard instruments, and Pythagorean tuning has been superseded by equal-tempered tuning. Nevertheless, the piano keyboard helps us to understand Pythagorean tuning. We can tune a piano in the Pythagorean manner by using the *sequence of fifths* shown in Figure 3.12. This sequence of fifths includes all 12 tones of the chromatic scale spread apart in different octaves. Starting with these 12 notes and proceeding by octaves, we can reach every note on the piano. We could begin by tuning the A in the middle of the keyboard (A4) to a standard pitch—for example, 440 Hz. Then we proceed by tuning the sequence in perfect fifths—upward E, B, F-sharp ..., and downward D, G, C.... All other notes on the piano are then tuned in octaves.

The flaw in this tuning system is that it makes the interval from G-sharp to E-flat so unpleasantly out of tune that it has earned the nickname *the wolf*. Nominally, this interval is a fifth, but Pythagorean tuning makes it

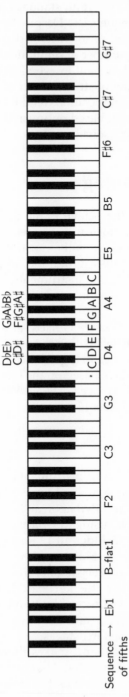

Sequence → E♭1   B-flat1   F2   C3   G3   D4   A4   E5   B5   F♯6   C♯7   G♯7
of fifths

D♭E♭   G♭A♭B♭
C♯D♯   F♯G♯A♯

C D E F G A B C

Figure 3.12. The piano keyboard. The sequence of fifths starts at E♭1 and proceeds up to G♯7. By setting these intervals as perfect fifths and tuning all other notes by octaves, we can achieve a Pythagorean tuning of the entire piano. This tuning was used in medieval polyphony until the thirteenth century. (See Schulter (1998).) Any seven contiguous notes in the sequence of fifths, define the tones in a major diatonic scale. For example, the tones F-C-G-D-A-E-B can be rearranged to the C-major diatonic scale C-D-E-F-G-A-B. Only six such scales are musically usable, corresponding to the major keys of B♭, F, C, G, D, and A. The problem with the remaining keys is that they permit the badly out-of-tune *wolf interval* E♭-G♯. Any five contiguous notes in the sequence of fifths defines the notes of a pentatonic (five-note) scale, used in Scottish and Chinese music.

50

about a quarter of a semitone[11] flat. The frequency ratio of this discrepancy is known as the *Pythagorean comma*.

If the piano keyboard had just a few more keys, we could proceed a fifth above G♯7 to D♯8. Traditionally, D♯ is the same as E♭. If we proceed downward seven octaves, we reach E♭1. But E♭1 has already been tuned by proceeding downward six fifths from A4. We might hope the these two ways of tuning E♭1 would be the same, but they are not. In fact, the ratio between these two tunings of E♭1 is equal to the Pythagorean comma. We can determine the value of the Pythagorean comma by the following computation. We proceed upward 12 fifths from E♭1 to D♯8. Then we return down seven octaves from D♯8 to a tone that is close to E♭1. In fact, it exceeds E♭1 by the Pythagorean comma. We can see this discrepancy by the following computation. The frequency ratio between the initial and final tones is

$$\left(\frac{3}{2}\right)^{12} \times \frac{1}{2^7} = \frac{3^{12}}{2^{19}} = \frac{531,441}{524,288} = 1.01364\ldots \tag{3.1}$$

Close, but not perfect! If we had returned to the same tuning of E♭1, this number would be 1 exactly. For comparison, the frequency ratio for an equal-tempered semitone is $1.05946\ldots$. Thus, the Pythagorean comma is about one-quarter of a semitone.

We specified (1) a starting pitch of E♭1 and (2) 12 consecutive upward fifths and then seven consecutive downward octaves. However, these two requirements are too restrictive. In fact, we can start with any pitch whatever, and we can mix upward fifths with downward octaves in any convenient manner so that we do not exceed the keyboard of the piano. In every case, the final pitch is above the initial pitch, and the frequency ratio between the two is equal to the Pythagorean comma.

Approximating m octaves with n fifths

Is there some way to modify this method, climbing by fifths and descending by octaves, so that we return to exactly the same pitch? No, it is not possible. If we climb n fifths and descend m octaves, the corresponding ratio is

$$\left(\frac{2}{3}\right)^n \times \frac{1}{2^m} = \frac{2^{n-m}}{3^n}$$

For this ratio to be equal to 1, we must have $3^n = 2^k$, where k denotes $n - m$. But this is impossible because the left side of this equality would be odd and the right side even.

We cannot climb by perfect fifths, descend by octaves, and return to exactly the same pitch. But is it possible to return closer than the Pythagorean comma? Suppose we ignore the difficulty and persist mixing upward

perfect fifths and downward octaves. Will we ever get closer to our starting pitch than the error of the Pythagorean comma? Curiously, the answer
is affirmative. For example, 53 upward fifths combined with 31 downward
octaves brings us to our original pitch with an error equal to about $1/7$ of
the Pythagorean comma. This can be seen by comparing the following
calculation with the Pythagorean comma as given in formula (3.1):

$$\left(\frac{3}{2}\right)^{53} \times \frac{1}{2^{31}} = \frac{3^{53}}{2^{84}} = \frac{19,383,245,667,680,019,896,796,723}{19,342,813,113,834,066,795,298,816}$$

$$= 1.00209\ldots$$

Ignoring the issues of practicality and musicality, the above fact could
lead to a scale that divides the octave into 53 tones instead of 12. Was the
number 53 merely a good guess? Or the result of an exhaustive search?
The answer is *no* to both questions because 53 was the result of a systematic computation. In fact, as we will see, this number was found by an
application of *back-and-forth subtraction* (BAFS), introduced in the preceding chapter.[12]

Instead of requiring $(3/2)^n = 2^m$, where n and m are the lengths of the
upward chain of fifths and the downward chain of octaves, we ask that
$(3/2)^n$ is *approximately* equal to 2^m. Taking logarithms, we want $n \log 3/2$ to
be close to $m \log 2$. Equivalently, we want

$$\frac{n}{m} \approx \frac{\log 2}{\log 3/2} = 1.70951\ldots$$

The preceding chapter discussed the ancient method of *BAFS*, which
is equivalent to the modern method of *simple continued fractions*. BAFS enables us to approximate an arbitrary ratio by a ratio of natural numbers.
Putting $\log 3/2 \approx 0.176091$ and $\log 2 \approx 0.301030$, the following is the calculation of the BAFS of $\log 2 : \log 3/2$.[13]

$$0.301030 - \mathbf{1} \times 0.176091 = 0.124939$$
$$0.176091 - \mathbf{1} \times 0.124939 = 0.051152$$
$$0.124939 - \mathbf{2} \times 0.051152 = 0.022635$$
$$0.051152 - \mathbf{2} \times 0.022635 = 0.005882$$
$$0.022635 - \mathbf{3} \times 0.005882 = 0.004989$$
$$0.005882 - \mathbf{1} \times 0.004989 = 0.000893$$

.   .   .

The bold numbers 1, 1, 2, 2, 3, 1,... are the partial quotients of this BAFS (or continued fraction). We write

$$\frac{\log 2}{\log 3/2} = [1; 1, 2, 2, 3, 1, \ldots] \tag{3.2}$$

The right side of this equation is an abbreviation for the infinite simple continued fraction

$$1 + \cfrac{1}{1 + \cfrac{1}{2 + \cfrac{1}{2 + \cfrac{1}{3 + \cfrac{1}{1 + \cfrac{1}{\ddots}}}}}} \tag{3.3}$$

The theory of continued fractions[14] asserts that by truncating this infinite continued fraction we obtain approximations of the left side of equation (3.2) by rational numbers (quotients of integers) that are in a certain sense the *best possible*. Specifically, we have the following approximations:

$$\frac{2}{1} = 1 + \frac{1}{1} \qquad\qquad \frac{5}{3} = 1 + \cfrac{1}{1 + \cfrac{1}{2}}$$

$$\frac{12}{7} = 1 + \cfrac{1}{1 + \cfrac{1}{2 + \cfrac{1}{2}}} \qquad\qquad \frac{41}{24} = 1 + \cfrac{1}{1 + \cfrac{1}{2 + \cfrac{1}{2 + \cfrac{1}{3}}}}$$

$$\frac{53}{31} = 1 + \cfrac{1}{1 + \cfrac{1}{2 + \cfrac{1}{2 + \cfrac{1}{3 + \cfrac{1}{1}}}}}$$

$$\cdot \qquad \cdot \qquad \cdot$$

The fractions 2/1, 5/3, 12/7, 41/24, and 53/31 are successively closer approximations of

$$\frac{\log 2}{\log 3/2}$$

Table 3.1. n fifths approximates m octaves. *Comma* denotes the frequency ratio between n fifths and m octaves. The Pythagorean comma 1.013643 is the second item in the third row. The fractions n/m are the *convergents* of the infinite continued fraction (3.3).

Fraction n/m	Comma $1.5^n/2^m$	Comma per fifth $1.5/2^{m/n}$	Percent error
$2/1$	$1.5^2/2 = 1.125000$	1.060660	$+6.0660\,\%$
$5/3$	$1.5^5/2^3 = 0.949219$	0.982778	$-1.7222\,\%$
$12/7$	$1.5^{12}/2^7 = 1.013643$	1.001130	$+0.1130\,\%$
$41/24$	$1.5^{41}/2^{24} = 0.988603$	0.999720	$-0.0280\,\%$
$53/31$	$1.5^{53}/2^{31} = 1.002090$	1.000039	$+0.0039\,\%$

Furthermore, these fractions have a remarkable musical interpretation. For each fraction n/m, the interval of n fifths is close to m octaves. Table 3 shows the level of approximation for each fraction.

Equal-tempered tuning

The dissonance of the interval E♭–G♯, the wolf interval, was an unwanted artifact of the Pythagorean system of tuning. The wolf's presence would make only six major and minor keys usable—those keys that do not include either of the pairs E♭, A♭ or D♯, G♯. This limitation on musical expression was not a problem until the end of the thirteenth century when musical taste became more complex.

Rather than continue to avoid the wolf, musicians developed schemes for *tempering* (i.e., adjusting) the Pythagorean scale. These tuning methods greatly enriched musical expression by permitting the free use of all keys. Accordingly, Bach composed a monumental cycle of 48 compositions, *The Well-Tempered Clavier, Books I & II*. Each *Book* contains 24 compositions: a prelude and fugue in each of the 12 major and 12 minor keys. However, Bach probably did not have access to *equal-tempered* tuning, the tuning system in universal use today.

It is said that Beethoven composed on pianos that were tuned by a system called *just* tuning, which gave each different key a different character. For example, when a piece was transposed from C major to D major, then, nominally at least, each note was raised a whole tone. However, the transposed piece acquired a different sound because the *just*-tuned D major scale had slightly different intervals than the C major scale. It has been suggested that to hear Beethoven's piano music as he intended it, we should hear it played on a piano with *just* tuning.

Equal-tempered tuning, also called *equal temperament*, is the only tempering of Pythagorean tuning in which transposition to a different key is achieved by raising (or lowering) each note exactly the same interval. Equal temperament is based on the fact, discussed above, that an interval of 12 fifths is quite close to seven octaves. If all fifths are slightly flattened by the same amount, then 12 consecutive fifths becomes exactly the same as seven octaves. Table 3 shows that each fifth needs to be reduced by 0.1130 %. The entire piano can be tuned by the pattern of Figure 3.12 using slightly flattened fifths. Equal temperament is equivalent to defining the semitone so that the frequency of the upper tone is always equal to the frequency of the lower tone multiplied by the factor $2^{1/12}$ (1.059463...). For example, if A is 440 Hz, then the equal-tempered frequency of A♯ is equal to $440 \times 1.05946 = 466.16$ Hz.

Table 3.2 compares the tuning of the chromatic scale from C4 to C5 according to the Pythagorean and equal-tempered methods. This version of the Pythagorean chromatic scale was used (possibly among others) in medieval polyphony. (See Schulter (1998).) Pythagorean tuning is based on the principle that all intervals can be obtained as a succession of fifths and octaves. For example, we can make a journey over the piano keyboard that starts at middle C and ends one semitone higher at C♯. First follow an ascending sequence of seven perfect fifths: C G D A E B F♯ C♯; then descend four octaves to the C♯ immediately above our initial C. To achieve an equivalent to this journey without exceeding the piano keyboard, one may mix the seven ascending fifths with the four descending octaves in any convenient order.

Table 3.2. Comparison of Pythagorean and equal-tempered tuning with A4 set at 440 Hz. A *cent* is one-hundredth of a semitone.

Note	Hz	Frequency ratio from C4	Frequency ratio from prev.	Cents from C4	Cents from prev.	Hz	Cents from C4	Cents from prev.
		Pythagorean temperament					**Equal temp.**	
C4	260.74	1/1		0.00	0.00	261.63	0	
C♯4	278.44	2187/2048	2187/2048	113.69	113.69	277.18	100	100
D4	293.33	9/8	256/243	203.91	90.22	293.66	200	100
E♭4	309.03	32/27	256/243	294.13	90.22	311.13	300	100
E4	330.00	81/64	2187/2048	407.82	113.69	329.63	400	100
F4	347.65	4/3	256/243	498.04	90.22	349.23	500	100
F♯4	371.25	729/512	2187/2048	611.73	113.69	369.99	600	100
G4	391.11	3/2	256/243	701.96	90.23	390.00	700	100
G♯4	417.66	6561/4096	2187/2048	815.64	113.68	415.30	800	100
A4	440.00	27/16	256/243	905.87	90.23	440.00	900	100
B♭4	463.54	16/9	256/243	996.09	90.22	466.16	1000	100
B4	495.00	243/128	2187/2048	1109.78	113.69	493.88	1100	100
C5	521.48	2/1	256/243	1200.00	90.22	523.25	1200	100

The fifths mentioned above are perfect fifths; that is, the frequency of the higher tone must be equal to the frequency of the lower tone multiplied by 3/2. An octave upward is obtained by doubling the frequency. Hence, each step of our keyboard journey corresponds to multiplication or division by 3/2 or 2, respectively. It follows that the frequency ratio of every Pythagorean interval is a ratio between a power of two and a power of three, as shown in the third and fourth columns of Table 3.2 — confirming the Pythagorean requirement that all intervals be associated with ratios of whole numbers.

The equal-tempered chromatic scale divides the octave into 12 equal semitones — equal in the sense that every semitone has the same frequency ratio, $2^{1/12} = 1.059463\ldots$. The equal-tempered semitone is divided into 100 equal parts called *cents*. The Pythagorean chromatic scale contained two different kinds of semitones: an interval of 90.22 cents, called a *diatonic semitone*, and a slightly larger interval of 113.69 cents, in medieval times known as an *apotome*.

In equal-tempered tuning, a fifth is tuned so that the frequency of the upper note is equal to the frequency of the lower note multiplied by $2^{7/12} = 1.4983\ldots$; the corresponding number for Pythagorean tuning is exactly 1.5. Octaves are the same in equal-tempered and Pythagorean tuning because $2^{12/12} = 2^1 = 2$.

With equal-tempered tuning, the sequence of fifths in Figure 3.12 becomes the *circle* of fifths because proceeding upward 12 equal-tempered fifths we span seven octaves *exactly* — not just approximately as in Pythagorean tuning.

To achieve equal-tempered tuning, a piano tuner starts with a perfect fifth and reduces this interval until *beats* (see page 46) at a certain rate are heard.[15]

The concept of equal-tempered tuning was known to the ancient Chinese. About 50 BCE, King-Fang, a Chinese scholar of the Han dynasty, found that it might be desirable to divide the octave into a scale of 53 equal intervals.[16] No doubt he found this number by exhaustive search. However, his recommendations are confirmed above by the fact that 53/31 is a convergent of the continued fraction (3.3). The microtonal scales dividing the octave into 53 equal intervals give truer fifths than the 12-tone equal-tempered chromatic scale.[17] In fact, the right column of Table 3 shows the precise improvement in the percentage error in the rendition of fifths.

We have traced the ideas of consonance and dissonance from the Pythagoreans to modern times. Musical intervals and their associated numerical ratios have lost their mystical significance, but they have acquired remarkable connections to diverse sciences.

I think that we have, in large measure, answered the questions posed on page 35. The preference for certain musical intervals is not entirely cultural. Dissonance and consonance — the foundation of melodic and harmonic tension and resolution—are based on the anatomy of the ear and the physics of sound. In various cultures, traditional musical structures have accommodated to these limitations through a process of natural selection—the survival of the musical. Musical innovators cannot expect to please audiences with music based on a purely mathematical system of composition that ignores the anatomical and acoustical basis of dissonance and consonance.

In the next two chapters, we look at geometric matters. Chapter 4 deals with curvature through the experiences of the inhabitants of a fantastic world called Tubeland.

Part II

The Shape of Things

Though there never were a circle or triangle in nature, the truths demonstrated by Euclid would for ever retain their certainty and evidence.

—DAVID HUME 1711–76,
An Enquiry Concerning
Human Understanding

4

Tubeland

*I call our world Flatland, not because we call it so, but to make its
nature clearer to you, my happy readers, who are privileged to live in
Space.*

*Imagine a vast sheet of paper on which straight Lines, Triangles,
Squares, Pentagons, Hexagons, and other figures, instead of remaining
fixed in their places, move freely about, on or in the surface, but without
the power of rising above or sinking below it, very much like
shadows—only hard and with luminous edges—and you will then have
a pretty correct notion of my country and countrymen. Alas, a few
years ago, I should have said "my universe": but now my mind has
been opened to higher views of things.*

—EDWIN ABBOTT ABBOTT, Flatland (1884)

IS THE SURFACE OF THE EARTH curved or flat? Despite the superficial
appearance of flatness, learned people as far back as Pythagoras in
the sixth century BCE have agreed that the surface of the earth is not
flat but spherical. In fact, the Alexandrian mathematician and astronomer
Eratosthenes of Cyrene (276?–195? BCE), gave a close approximation of the
circumference of the earth using measurements of the shadows of objects
of known height at different latitudes. It is a myth—probably initiated by
Washington Irving in his *History of Christopher Columbus* (1828)—that in
the time of Columbus there was a widely held belief in a flat earth.

Today, we can look down on the earth from miles above, and we can
see that its surface is not flat but spherical. In the fantasy *Flatland* (Abbott (1884)), the narrator, A. Square, and the other characters (triangles,
squares, etc.) cannot observe their world in a similar fashion because
they cannot leave the surface in which they live. There is a sequel to Abbott's fantasy[1] in which Flatlanders adopt the theory that they live on a
sphere—not on a plane. We will see that there is a way for them to settle
this question, even if they are unable to circumnavigate their sphere.

A difference between us and the Flatlanders is that we can observe, to a limited fashion, how the surface of the earth is embedded in space. Geometers would say the Flatlanders can only observe the *intrinsic geometry* of their surface, but, on the other hand, we are able to observe the *extrinsic geometry* of the earth's surface. We can leave the earth's surface, but we can never leave the universe. Consequently, we can only learn the *intrinsic* geometry of the universe.

For thousands of years it was supposed that Euclidean geometry provided the only possible geometry for the universe. In the first decades of the twentieth century, Euclid's long reign came to an end with the advent of Albert Einstein's theory of relativity. Einstein devised a geometry of four dimensions — one temporal and three spatial dimensions — that explained puzzling questions of physics.

The quest to understand the geometry of our universe continues today. In April and May 2000, scientists claimed that observations of cosmic microwave background radiation from two research balloons, BOOMERANG[2] and MAXIMA,[3] support the conclusion that the universe is flat.

The geometry of the cosmos — the cosmological questions of curvature and flatness — would become much simpler if we only lived in a universe of lower dimensionality. This chapter is an inquiry into curvature and flatness in at most two dimensions. Curves and surfaces in lower dimensions set the stage for higher dimensional cosmological questions, but also they have, in their own right, a rich complexity. We begin with one-dimensional objects — straight lines and curves.

Curvature of Smooth Curves

We discuss only curves that are *smooth*. In Part IV we will examine more critically the concept of smoothness, but for now we can imagine a length of wire in the shape of the curve. For these smooth curves, we assume that it makes sense to speak of the arc length between two points on the curve.

The inner world of the curve-bound inchworm

The following story has no basis in natural history, but it might clarify the concept of *inner* or *intrinsic* geometry. Imagine that an inchworm lives on a curve and in accordance with the following rules:

1. He is unable to leave his curve.
2. He can move along his curve and measure distances.
3. He is blind and has no sense of up or down or movement. Measuring distances is his only perception of the world.

What the inchworm perceives is the *inner* or *intrinsic* geometry of his curve. The boring triviality of his one-dimensional world is evident. He can go backward and forward and measure how far he has gone, but he is unable to perceive, for example, the shape of his curve. There is little else to say about this poor creature.

However, we will see that a similar inchworm that lives on a 2-dimensional *surface* can obtain much more complex information about his world.

To find interesting information about a curve, we must examine how it is embedded in two or three dimensions — that is, we must look at the *extrinsic* geometry of the curve. We begin with the two-dimensional case.

Curves embedded in two dimensions

The circle is the standard of comparison for the curvature of smooth curves in a fixed plane. There is a precise quantitative definition of *curvature of a curve*. The *curvature* of a circle of radius R is defined as $1/R$. The curvature of an arbitrary smooth curve at a particular point P is the curvature of the circle — called the *osculating circle* — that fits the curve best at point P. Examples of osculating circles are shown in Figure 4.1.

A curve-bound inchworm has no means of detecting the curvature of his curve. In fact, the curvature of a curve is an extrinsic property of the curve.

In Figure 4.1(b), the osculating circles C and D are on opposite sides of the curve, corresponding to opposite *senses of concavity* of the curve — concave up and concave down. For plane curves, the curvature is usually given a plus or minus sign, arbitrarily, to distinguish the two possible senses of concavity — usually concave-up curvature is given a positive sign. For example, in Figure 4.1(b), we could give the curvature a plus sign at S and a negative sign at U.

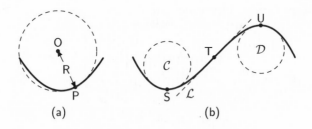

(a)   (b)

Figure 4.1. Osculating circles.

(a) The osculating circle touches the solid curve at point P on the solid curve. The center O of this circle is called the *center of curvature*; the radius R is called the *radius of curvature*; and 1/R is called the *curvature* at point P.

(b) Osculating circles C and D touch the solid curve at points S and U, respectively. At point T, the osculating circle degenerates to the line L, and the curvature is 0.

For a straight line there is no best-fitting circle. Larger circles fit better than smaller ones. The curvature of a straight line is 0. However, it is possible for a curve to have 0 curvature at a single point without being a straight line. In fact, in Figure 4.1(b) the curvature at T is equal to 0. The osculating circle at T degenerates to the straight line $\mathcal{L}$. If the point S slides down (or if the point U slides up) and approaches point T, the center of the osculating circle tends — one could say it explodes — to infinity.

In two dimensions, the (signed) curvature of a curve completely determines its shape. If we specify a certain curvature at each point on a piece of thread, then there is only one way to shape the thread into a plane curve with the specified curvature at each point.

Curves embedded in three dimensions

The helix

In two dimensions, circles are a sufficient paradigm for the local behavior of curves, but this is not the case for curves embedded in three-dimensional space. The simplest example that shows the essential features of curves in three dimensions is the thread of a cylindrical screw, shown in Figure 4.2(a) — also the familiar shape of a cylindrical steel spring. The mathematical name for this curve is *helix*. (The famous double helix of molecular biology is an example with much additional structure.) Archimedes (287?–212 BCE) invented a water pump using a helical screw.

Osculating circles are defined for smooth curves in three dimensions. The curvature at a point on a curve is defined as the (nonnegative) curvature of the osculating circle. The plane of the osculating circle is tangent to the curve. However, in three dimensions, the curvature of the oscu-

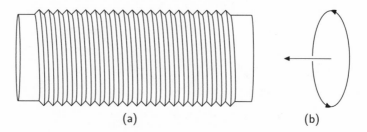

(a)   (b)

Figure 4.2. Cylindrical right-handed screw. (a) The curve formed by the thread of a cylindrical screw is a helix. (b) When rotated clockwise, the right-handed screw advances to the left.

Right-hand rule. Grasp the screw in the right hand with the fingers pointing in the direction of rotation. Then the screw advances in the direction pointed by the thumb.

lating circle is not sufficient to describe the local behavior of a curve. A second quantity called *torsion* is defined at each point of the curve. The torsion is a measure of the local tendency of the curve to leave its tangent plane. By convention, the torsion of a helix is positive or negative depending on whether it represents the thread of a right- or left-handed screw. The tendrils of grape and hops vines have positive and negative torsion, respectively.

The pitch of a screw is the distance that the screw advances with a turn of one revolution. Suppose that a right-handed cylindrical screw has radius a and pitch p. Putting $b = p/(2\pi)$, the curvature κ and the torsion τ are given by the following formulas:

$$\kappa = \frac{a}{a^2 + b^2} \qquad \tau = \frac{b}{a^2 + b^2}$$

From these formulas, we see that if the pitch is fixed, torsion can be made as small as desired for the radius a sufficiently large. Furthermore, torsion becomes as large as desired if both the radius a and the pitch p are sufficiently small.

Curvature of Smooth Surfaces

Surfaces like the plane, the sphere, and the torus are smooth. A surface is smooth if the following two conditions are met:

1. Every point P on the surface has a tangent plane.
2. Slight changes in P correspond to only slight changes in the tangent plane.

The next section defines curvature of a surface using the curvature of curves contained on the surface. One might think that curvature of a surface, like curvature of a curve, is an extrinsic concept, but, surprisingly, it can be shown that it is intrinsic.

In analogy to the curve-bound inchworms (pages 62–63), we can imagine surface-bound inchworms who live on a surface and measure distances there. We will suppose that surface-bound inchworms always measure the shortest path contained entirely on their surface. (These shortest paths are called *geodesics*.) The surface-bound inchworms are also able to measure the angle between two geodesics. In short, the surface-bound inchworms have access to the intrinsic geometry of their surface. The curve-bound inchworms have a much less interesting existence than their surface-bound cousins because it turns out that the latter are able to measure the curvature of their surface — an ability that seems all the more surprising because the curvature of surfaces is defined below in a way that seems to be extrinsic and hence outside of the purview of inchworms. (Recall that curve-bound inchworms are *not* able to measure the curvature of the curve on which they live.)

The denizens of Flatland–Sphereland in the fantasies of Abbott and Burger are constrained to make observations only of the intrinsic geometry of their world. They are, in fact, a species of surface-bound inchworms. We will see more details in the discussion of Tubeland below.

The curvature of surfaces was first defined by Carl Friedrich Gauss (1777–1855) and is generally known as *Gaussian curvature*. Gauss's definition, which we will see in the next section, makes use of the manner in which a surface is embedded in three-dimensional space; in other words, it uses *extrinsic* properties of a surface. Gauss discovered that his curvature only *seems* to be an extrinsic property. Although the definition of curvature is *ex*trinsic, nevertheless the Gaussian curvature of a surface is an *in*trinsic property of the surface. This is why surface-bound inchworms can determine the curvature of their surface without ever leaving it. Although Gauss was generally sparing in his use of superlatives, he extolled this discovery, calling it the *theorema egregium*, the extraordinary theorem.

Gaussian curvature — Extrinsic definition

The definition of Gaussian curvature is illustrated in Figure 4.3 for two types of surfaces. Figures 4.3(a) and (b) show, respectively, a convex-concave[4] surface and a saddle surface. Since the construction is the same for both surfaces, the two figures contain the same labels. Point P is chosen arbitrarily on the surface. The directed line segment **n** is normal (perpendicular) to the surface. Each plane containing **n** intersects the surface in a curve. At P, one of these curves $\mathcal{K}$ achieves maximum curvature κ, and another curve $\mathcal{L}$ achieves the minimum curvature λ. These curvatures, κ and λ, are called the *principal curvatures* at point P. If the principal curvatures are unequal, then it can be shown that the angle of intersection

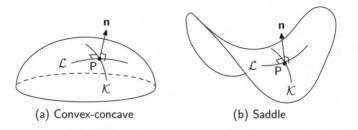

(a) Convex-concave   (b) Saddle

Figure 4.3. Curvature of surfaces. In both (a) and (b), **n** is normal to the surface at P. Curves $\mathcal{K}$ and $\mathcal{L}$ are formed by the intersection of planes through **n** with the largest and smallest possible curvatures at P. The product of these two curvatures is the Gaussian curvature at P.

of the curves $\mathcal{K}$ and $\mathcal{L}$ is 90°, as shown in Figure 4.3. For the saddle point in Figure 4.3(b), one of the principal curvatures is given a positive and the other a negative algebraic sign. Whatever the sign choice, the curvatures of κ and λ have opposite signs. *The Gaussian curvature at P is defined as the product $\kappa\lambda$ of the two principal curvatures.* The curvature is positive at a convex-concave point (Figure 4.3(a)) and negative at a saddle point (Figure 4.3(b)).

The mathematical terms for convex-concave and saddle points are, respectively, *elliptic* and *hyperbolic* points. Points where exactly one or both of the principal curvatures are 0 are called, respectively, *parabolic* or *planar* points. The lateral surface of a circular cylinder consists of parabolic points, and a plane consists of planar points. However, a planar point can be found on a surface that is not a plane. In fact, point P in Figure 4.4 is a planar point.

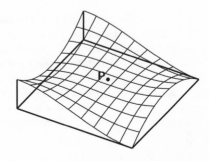

Figure 4.4. Monkey saddle. The point P is the only anomalous point. Every other point on this surface is an ordinary saddle point.

The surface shown in Figure 4.4 is a *monkey saddle*—so named because a monkey needs a place, not only for his two legs, but also for his tail. At point P, the monkey saddle point, the principal curvatures are both 0, and, therefore, the Gaussian curvature is also 0. Every other point of the surface is an ordinary saddle point with negative Gaussian curvature.

On a plane, the Gaussian curvature is 0 everywhere. If the plane is rolled up into a scroll, the curvature remains 0. On a sphere, the curvature has a positive constant value equal to $1/R^2$ where R is the radius of the sphere. The sphere is the only surface with constant positive curvature. Physically, this means that the *sphere*, unlike the plane, cannot be "bent" without stretching. This is why a spherical shell is more rigid than a rectangular box. A surface that has constant *negative* curvature is less familiar. A surface of this sort, known as a pseudosphere, is shown in Figure 4.5.

Every point of the pseudosphere is a saddle point. The principal curvatures κ and λ are achieved by (1) vertical circular cross-sections, and (2) the intersection with the surface of planes through the horizontal axis. It is plausible that the Gaussian curvature of the pseudosphere in Figure 4.5(a) is constant because as the magnitude of κ increases (i.e., as the vertical circular cross-section becomes a smaller circle), the corresponding magnitude of λ decreases (i.e., radius of the osculating circle of the radial cross-section increases). The pseudosphere (Figure 4.5(a)) is swept out by rotating a curve called a tractrix about its horizontal axis. In Figure 4.5(b), the curve is rotated about the base of the enclosing rectangle.

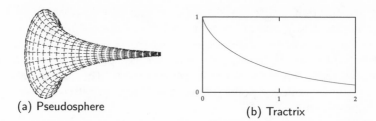

(a) Pseudosphere

(b) Tractrix

Figure 4.5. The pseudosphere is generated by rotating a tractrix. The pseudosphere (a) is swept out by rotating the tractrix (b) about the horizontal axis. The tractrix can be continued indefinitely to the right. The tractrix is the curve described by dragging a weight initially at the point (0,1), the upper left corner of the enclosing rectangle, along a horizontal surface by means of a rope 1 unit in length. The end of the rope is initially at (0,0). The rope is pulled to the right along the bottom of the enclosing rectangle. The curve does not depend on the magnitude of the frictional resistance — the surface could equally well be ice or concrete.

The intrinsic nature of Gaussian curvature plays a central role in the following sequel to *Flatland* (Abbott (1884)) and *Sphereland* (Burger (1965)). The Flatlanders fall short of actually proving the *theorema egregium*, but they make some observations that confirm Gauss's result.

Tubeland — A fantasy

Certain further observations by the Flatlanders have shown that their universe is neither a plane nor a sphere. In fact, Flatlanders now claim to live on a two-dimensional surface like an automobile inner tube — known to mathematicians as a *torus* or *anchor ring*, shown in Figure 4.6. The following is an account of that discovery.

Abbott wrote *Flatland* partly as a satire — directed, for example, at the Victorian treatment of women. We limit ourselves to the discussion of the geometry of Tubeland (also known as Flatland or Sphereland). Let us review the basic geometric facts of life in Tubeland–Sphereland–Flatland:

- Tubelanders are two-dimensional beings. Since they cannot leave their two-dimensional world to examine it from a three-dimensional

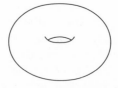

Figure 4.6. A torus: the world of Tubeland.

vantage point, there is no direct way for them to perceive that their universe is a torus. They must study their universe as we do ours: by devising and testing hypotheses based on scientific investigation.

- In Tubeland, a line segment is defined as the shortest path connecting its two endpoints *lying entirely on the surface of the torus*. Tubelanders persist in this terminology although mathematicians would prefer the term *geodesic* because, in general, these paths are not straight lines in the usual three-dimensional sense.
- Tubelanders can to draw and measure "line segments" and angles.

The Sphereland hypothesis was based mainly on two experiments:

Experiment 1. *Circumnavigation:* Two Flatland Columbuses made extended journeys in opposite directions and, after various adventures, met in a distant region.

Experiment 2. *Triangles:* It was discovered that the sum of the interior angles of a certain triangle was greater than $180°$—a discrepancy that could not be explained by experimental error.

Of these two, the triangles experiment is the most interesting, but first we give a critique of the circumnavigation experiment.

Experiment 1 shows that Flatland is not a plane. However, circumnavigation is consistent with the geometry of a torus. In fact, examining the figure at the beginning of this section, it is easy to see that a torus permits two different types of circuits: a large circuit in a horizontal plane or a smaller circuit in a vertical plane. Furthermore, circumnavigation is consistent with other hypotheses concerning the geometry of Flatland.

The triangle surveys

Flatlanders rediscovered Euclidean geometry and long considered it the definitive account of the geometry of their world. The starting point for Experiment 2, the triangles experiment, is a theorem from Euclidean plane geometry.

Figure 4.7.

Theorem 4.1. *The sum of the three internal angles of a triangle is equal to a straight angle.*[5] For example, in Figure 4.7, $\alpha + \beta + \gamma = 180°$.

After the successful circumnavigation, Flatland scientists were certain that their world was not flat, not a Euclidean plane. Furthermore, they were now certain that some of the theorems of Euclidean geometry must be false. A leading scientist, Azimuth Aphelion, believed that Flatland might be a very large sphere and recommended testing Theorem 4.1. He claimed that although this theorem had been repeatedly verified, within experimental error, by draftsmen and surveyors, perhaps a discrepancy would show up for very large triangles. The angles of triangle S (Figure 4.8) were carefully measured, and the sum was found to *exceed 180°*, contrary to Theorem 4.1.

This discovery created a great stir in Flatland scientific circles. A consensus emerged that this experiment showed that Flatland could not possibly be flat, and most believed that Aphelion's Sphereland hypothesis was adequately established.

Of course, there were some doubters. The chief of them was Peridot Perigee, who organized an expedition to check the Sphereland hypothesis by measuring a second triangle T in a different part of Flatland. To the astonishment of the scientific world, Perigee found that the sum of the angles of T was significantly less than 180°. Perigee noted that her discovery was inconsistent with the Sphereland hypothesis because on a sphere this sum must be *greater* than 180°.

Epilogue

Perigee was on the point of announcing her Tubeland theory, when it seemed that a great disaster overtook Flatland–Sphereland–Tubeland. The Mechanic deflated the tube and threw it on a disorderly pile of other deflated tubes.

This event was certainly an indignity, but was only a *seeming* disaster for Tubeland. In fact, life in Tubeland continued with very little change because the geometry of Tubeland was changed only in subtle ways. Distances were slightly reduced, and geodesics were altered only minutely.

The geometry of a surface that deals only with the measurement of geodesic lengths *on the surface* is called the *inner* or *intrinsic* geometry of the surface. The inner geometry of a two-dimensional surface contains no information concerning the manner in which the surface is *embedded* in three-dimensional space. *Bending* a surface, without stretching or com-

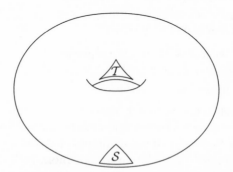

Figure 4.8. Flatlanders measured the sum of the interior angles of the two triangles. For S the sum exceeded 180°, and for T it was less than 180°.

pressing it, has no effect on its inner geometry. For example, the inner geometry of a plane does not detect the difference between a ordinary plane and one that has been rolled up into a scroll.

The triangle survey projects were an attempt to understand the inner geometry of Tubeland.

When we see the Mechanic deflate the tube and throw it on a pile, we see a large change in the way in which the tube is embedded in 3-dimensional space, but we see little change in the inner geometry of the tube.

For the Tubelanders, the manner of embedding their tube in 3-dimensional space has no meaning. Similarly, we can never hope to learn more than the inner geometry of our universe.

Triangular excess

The triangle surveys of the Flatlanders merit further comment. What can be discovered in general by this type of measurement? Consider first a triangle survey on a sphere.

A triangle bounded by great circles on a sphere, for example, T in Figure 4.9, is called a *spherical triangle*. The sum $\alpha + \beta + \gamma$ of the interior angles of a spherical triangle is always greater than 180°. (Recall that 180° is equal to π radians.) The difference measured in radians, that is,

$$\alpha + \beta + \gamma - \pi \qquad (4.1)$$

is called the *spherical excess*.

Let the radius of the sphere be R. Then the surface area of the spherical triangle T is equal to the spherical excess multiplied by R^2; that is, the area is equal to

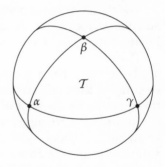

Figure 4.9. Spherical triangle.

$$R^2(\alpha + \beta + \gamma - \pi)$$

The above is known as Girard's formula, after French mathematician Albert Girard (1595–1632).

For a triangle drawn on a surface that is not a sphere (e.g., S or T in Figure 4.8), formula (4.1) is meaningful. However, if the surface is not a sphere, it is no longer appropriate to call formula (4.1) the *spherical* excess — *triangular excess* might be a better term — and the connection of formula (4.1) with surface area disappears. In Figure 4.8, the triangular excesses of S and T are positive and negative, respectively. Triangular excess is an intrinsic property because the angle between geodesics is intrinsic.

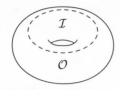

Figure 4.10. The region $\mathcal{O}$ is convex-concave, and region $\mathcal{I}$ consists of saddle points.

Triangular excess has a remarkable property that we illustrate using the torus of Figure 4.10. In this figure, the dashed circle represents the contact of a plane tangent with the top of the torus. There is a similar circle, not shown, where the torus might rest on a flat floor. These two circles divide the torus into two regions, the outside $\mathcal{O}$ and the inside $\mathcal{I}$. For an automobile inner tube, the outside is the part of the tube that makes contact with the tire tread. Any triangular region that lies entirely on the outside $\mathcal{O}$ has positive triangular excess, and any triangular region that lies entirely on the inside $\mathcal{I}$ has negative triangular excess. The subregions $\mathcal{O}$ and $\mathcal{I}$ have geometric properties that can be seen with the eye and felt with the hand. The region $\mathcal{O}$ is convex-convex, and $\mathcal{I}$ has everywhere the shape of a saddle (see Figures 4.11(a) and (b)). The regions $\mathcal{O}$ and $\mathcal{I}$ consist, respectively, of *elliptic* and *hyperbolic* points. On the boundary circles that separate $\mathcal{O}$ and $\mathcal{I}$, the surface is *planar*.

The classification of points on a smooth surface into hyperbolic, elliptic, and planar does not cover all possibilities, but these are the ordinary cases. Points that do not fit in these categories are called singularities.

The numerical value of the triangular excess—more than merely the algebraic sign—has a geometric meaning. Triangular excess turns out to be equal to the *total Gaussian curvature* of the triangular region. Dividing this quantity by the area of the triangular region is the *average* Gaussian curvature. Since triangular excess can be measured by Flatlanders, it must be an intrinsic property of surfaces, and, therefore, average Gaussian curvature must also be intrinsic. This is a confirmation of Gauss's *theorema egregium*, the theorem that asserts that Gaussian curvature is an intrinsic property of surfaces.

When a surface-bound inchworm crawls on a surface, hyperbolic points look like mountain passes and elliptic points look like peaks or pits. He

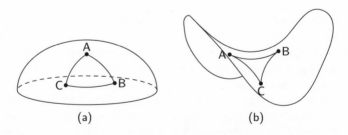

Figure 4.11. (a) A convex-concave surface. (b) A saddle surface.

is able to distinguish between hyperbolic and elliptic *even if he is extremely near-sighted*. In other words, this distinction depends only on *local* measurement of the surface.

The foregoing sections have discussed curvature and flatness as *local* properties. The next section discusses Euclidean geometry, which gives flatness a *global* meaning. Later we will see that *non*-Euclidean geometry gives curvature a global meaning.

Euclidean Geometry

The practical needs of construction and the division of land led the ancient Egyptians and Babylonians to consider certain geometric problems. For example, the Egyptians knew how to find the volume of a truncated pyramid with a square base, shown in Figure 4.12.[6] On the other hand, the Egyptians used an *incorrect* method for finding the area of a quadrilateral.[7] Clearly, the Egyptians lacked a rigorous method of deriving and testing their geometric assertions.

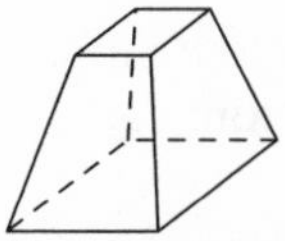

Figure 4.12. Truncated pyramid.

However, in the sixth century BCE the Greeks made a crucial advance in the history of science. They discovered that geometric facts can be deduced from a small number of self-evident assumptions called axioms.[8] This systematic method of proof is called the *axiomatic method*, an intellectual tool of continuing importance in mathematics and the sciences. In *The Elements*, Euclid (fl. 300 BCE) compiled geometry as a formal axiomatic system. *The Elements* contains the work not just of Euclid alone, but of many ancient Greek mathematicians. This book is still the primary source worldwide for school geometry—the all-time longest lasting academic textbook.

We become convinced of geometric facts in many different ways—by looking at a figure, by making measurements, or by accepting the authority of a teacher or a book. Euclidean geometry seeks not so much to convince us as to show us how geometric facts can be derived logically from a small number of assumptions called axioms. This method of proof, the axiomatic method, is systematic, consistent, shared, and reliable. In geometry and elsewhere, we may obtain brilliant ideas through a flash of insight or a flight of fancy. The axiomatic method enables us to give validity to our ideas—to put foundations under our "castles in the air."

A geometric proof often becomes more interesting if we focus on the *logic* of the proof—especially if we feel that we "know" the truth of the assertion to be proved. For example, Book I of *The Elements* contains the following:

Proposition 15. *If two straight lines cut one another, they make the vertical angles equal to one another.*

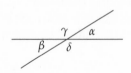

Figure 4.13. Vertical angles.

In Figure 4.13, the proposition asserts, for example, that angle α is equal to angle β. If we expect to find an exciting new fact here, we might be disappointed because our geometric intuition already tells us that this assertion is true. On the other hand, to show that this result follows from the axioms and previously proved results takes a degree of cleverness.

Proof. We note that $\alpha + \gamma = \gamma + \beta$ because each of these two sums is equal to a straight angle (180°). Since equals subtracted from equals are equal, it follows that $\alpha = \beta$. □

The *axiomatic method* is a refinement of the methods that we use in ordinary discourse when we argue by giving *reasons*. However, the axiomatic method is more rigorous because (1) it uses terms of greater clarity, (2) it requires logical rules of inference, (3) and it insists that every assertion be connected to a chain of inferences leading back to a small number of assumptions.

It has been said, facetiously, that geometry is the science of drawing correct conclusions from incorrectly drawn diagrams. Indeed, the chalkboard drawings presented in a classroom are not the true objects of geometric study. Euclidean geometry requires that diagrams be used merely to assist in the visualization of abstract geometric ideas. A diagram may elicit an insight of mathematical truth, but we are never allowed to jump to a conclusion merely by looking at a diagram. All inferences must follow logically from the axioms and from previously proved results.[9]

To avoid circularity, geometry—indeed, every axiomatic system—must have fundamental terms that are undefined. In Euclidean geometry, these undefined terms include *points, lines,* and *planes.*[10] They are presumed objects of ordinary experience. For example, a line may be a mark drawn in the sand, but for the purposes of cosmology *a line is the path followed by a light ray.*

In order to check the validity of our reasoning, it may help to ignore the meanings of the basic terms. This point of view was put forward by German mathematician David Hilbert (1862–1943), leader of the formalist movement in the theory of axiomatic systems. Hilbert once remarked, "One must be able to say at all times—instead of points, straight lines, and planes—tables, chairs, and beer mugs."

No experiment can confirm the logical correctness of Euclidean geometry. In fact, no one doubts the logical correctness of Euclidean geometry. To say that Euclidean geometry is logically correct means that valid application of the laws of inference implies certain conclusions — the theorems — from certain assumptions — the axioms.

The parallels axiom

There is only one controversial geometric axiom. There is no controversy over the following axiom.

Axiom 4.1 (incidence axiom). *Two distinct points determine exactly one line.*

In fact, the only axiom that is controversial is the *parallels axiom*, an axiom of plane, two-dimensional, Euclidean geometry.

Axiom 4.2 (parallels axiom). *Given any straight line $\mathcal{L}$ and a point P not on it, there exists one and only one straight line, coplanar with $\mathcal{L}$, that passes through the point P and never intersects the line $\mathcal{L}$, no matter how far the lines are extended.*[11]

Axiom 4.2 has a speculative quality not found in the other axioms of Euclidean geometry. The phrase "no matter how far the lines are extended" jumps to a conclusion because we cannot observe the entire infinite extent of a straight line. In fact, Axiom 4.2 has long been considered out of place in the canon of Euclidean axioms. For more than a thousand years, from Ptolemy (second century CE) to Adrien Marie Legendre (1752–1833), mathematicians tried, without success, to prove this axiom from the other axioms of Euclidean geometry. Their efforts were in vain because it is now known that no such proof exists. Gerolamo Saccheri (1667–1733), Jesuit priest and professor at the University of Pavia, attempted a very elaborate *indirect proof*. He assumed the falsity of the Axiom 4.2 and derived a large body of results. He believed, erroneously, that he had *proved* the parallels axiom. In 1733, he published his work under the title *Euclides ab Omni Naevo Vindicatus* (Euclid Freed of All Blemish). He failed to realize that his results were theorems in a new *non-Euclidean geometry*. Like Columbus, he failed to understand the identity of the vast new land he discovered.

Non-Euclidean Geometry

In the first half the nineteenth century, the time was ripe for the development of non-Euclidean geometry. Two mathematicians, the Russian Nicolai Lobachevsky (1792–1856) and the Hungarian János Bolyai (1802–60), independently pursued programs very similar to that of Saccheri, but with a different rationale. They were not trying to give an indirect proof of the parallels axiom. Rather, they examined a modification of Euclidean plane geometry: They replaced the parallels axiom with the axiom that given a line and point not on the line, there exist, not just one, but many lines containing the given point and not meeting the given line. Lobachevsky's results were published in 1829–30 and Bolyai's in 1832–33. On hearing of this work, Carl Friedrich Gauss, then a towering figure in the world of mathematics, initiated an unpleasant quarrel over priority by claiming that he made unpublished investigations of this sort as early as 1813.

Theorem 4.1 on page 69 is important here because the parallels axiom is used in its derivation. If one suspects—as did the Tubelanders—that the parallels axiom fails in the real world, one might test the triangle theorem experimentally, for example, by drawing triangles on paper and measuring the angles with care. This experiment has two difficulties: (1) In the cosmological sense, lines must be light rays, not lines drawn on paper. (2) A proper test probably needs to be done with a much larger triangle.

Gauss believed that the universe might be governed by non-Euclidean geometry—might not be "flat"—and he conducted an experiment to test this hypothesis. He measured the angles of a triangle formed by three mountain peaks. Gauss's triangle was formed by light rays, the standard cosmological straight lines. His experiment did not support the non-Euclidean hypothesis—he did not show a discrepancy with the Euclidean triangle theorem. He found that the sum of angles was 180° within experimental error. Some cosmologists—nonbelievers in a flat universe—may feel that Gauss's experiment was on the right track but that his triangle was much too small.

Non-Euclidean geometry is more than merely the denial of Euclidean geometry. There are two species of non-Euclidean geometry—more specifically, non-Euclidean versions of *plane* Euclidean geometry. They are based on two different substitutes for the parallels axiom (Axiom 4.2).

Axiom 4.3 (elliptic substitute for the parallels axiom). *Given a line $\mathcal{L}$ and point* P *not on* $\mathcal{L}$, *every line, coplanar with the line $\mathcal{L}$ and containing the point* P, *meets $\mathcal{L}$ in exactly one point.*

Axiom 4.4 (hyperbolic substitute for the parallels axiom). *Given a line $\mathcal{L}$ and point* P *not on* $\mathcal{L}$, *there exists more than one line, coplanar with the line $\mathcal{L}$, containing the point* P, *that does not meet $\mathcal{L}$.*

Because these are axioms of *plane* non-Euclidean geometry, these axioms do not need to assert that the lines under discussion are coplanar.

Axioms 4.3 and 4.4 lead to two different kinds of non-Euclidean geometry—*elliptic* and *hyperbolic*, respectively.

Bernhard Riemann (1826–66) pioneered elliptic geometry. Earlier, Saccheri, Lobachevsky, Bolyai, Gauss, and others created a body of theorems of hyperbolic geometry. Even though they proved many theorems of non-Euclidean geometry without any contradictory conclusions, how can we be sure that there will never be contradictions? The answer is that Euclidean geometry itself is able to demonstrate the validity of these non-

Euclidean geometries. To achieve this demonstration, we must organize a geometric masquerade in which Euclidean constructions mimic a non-Euclidean world.

Models of non-Euclidean geometries

We will see that non-Euclidean geometries can be realized within Euclidean geometry. This will give us a much greater confidence in non-Euclidean geometry because we will see that it cannot contain contradictions unless ordinary Euclidean geometry also has contradictions. Euclidean geometry has been around a long time, and most people believe that it does not contain contradictions.

To carry out this program, we must do the following:

1. Redefine the points and lines of non-Euclidean geometry as certain objects of Euclidean geometry. The newly defined "points" and "lines" need not be points and lines in the old sense, but they must be legitimate constructions of Euclidean geometry — not tables, chairs, or beer mugs.

2. Redefine, as necessary, basic concepts of Euclidean geometry, for example, "congruence." (Recall that in Euclidean geometry two figures are congruent if a *rigid motion* can bring the one into coincidence with the other.)

3. Show that the newly defined "points" and "lines" satisfy either Axiom 4.3 (elliptic) or Axiom 4.4 (hyperbolic). Additionally, excluding the parallels axiom, it must be shown that all the other axioms of Euclidean geometry are true.

A model of elliptic geometry

Our universe for elliptic geometry is the surface of a fixed sphere, as shown in Figure 4.14. We define "points" as pairs of antipodal points on the sphere (e.g., P and P'). "Lines" are great circles on the sphere, for example, the great circle C.

Two figures are "congruent" if the one figure can be brought in coincidence with the other by rotating the sphere about an axis through the center O of the sphere, or by a sequence of several such rotations. By *figure*, we mean a collection of "points" (pairs of antipodal points) and "lines" (great circles).

Axiom 4.4, the elliptic substitute for the parallels axiom, holds for this spherical model because every pair of great circles must intersect. There are no "parallel lines."

As seen in Figure 4.9, in elliptic geometry the sum of the angles of a triangle is always *greater* than 180°.

It is possible to verify the other axioms of Euclidean geometry — excluding the parallels axiom. This can be done, but it is a rather exacting task.

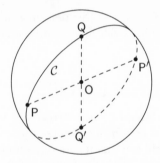

Figure 4.14. Spherical model for elliptic geometry.

We will verify just one axiom, the incidence axiom (Axiom 4.1): *Two distinct points determine exactly one line.* Suppose we have two distinct "points," two distinct pairs of antipodal points (P,P') and (Q,Q'). Since the pairs are distinct, the two points P and Q cannot be antipodal points and they cannot be identical. Therefore, there is exactly one great circle C connecting P and Q — the unique "line" connecting the two given "points." The great circle C also contains the points P' and Q', the antipodal points of P and Q, respectively. (Note that we have avoided the difficulty that there is more than one great circle — in fact, there are infinitely many — connecting a pair of antipodal points.)

A model of hyperbolic geometry

We will discuss a model of hyperbolic geometry that was introduced by the French mathematician Jules Henri Poincaré (1854–1912). In Poincaré's model, the "points" are ordinary Euclidean points in the interior of a fixed circle C. We exclude points on the circle or exterior to it. As shown in Figure 4.15, hyperbolic "lines" are arcs of circles that are orthogonal to the circle C, that is, that meet C at an angle of 90°.

Axiom 4.4, the hyperbolic substitute for the parallels axiom, is illustrated in Figure 4.15(b). Given the hyperbolic line $\mathcal{H}$ and a point P that does not lie on $\mathcal{H}$, there exists more that just one hyperbolic line containing P that does not meet $\mathcal{H}$. The figure shows two such lines, but there are infinitely many.

The incidence axiom (Axiom 4.1) holds in Poincaré's model because in Euclidean plane geometry it can be shown that given two points P and Q and a circle C there exists one and only one circle orthogonal to C containing the points P and Q.

Points in Poincaré's model are essentially Euclidean points. In this sense, this hyperbolic model is more straightforward than the elliptic model in which a non-Euclidean point is a pair of antipodal points on a sphere.

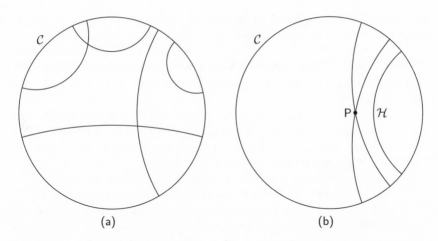

(a)   (b)

Figure 4.15. Poincaré's model of hyperbolic geometry. Hyperbolic lines are circular arcs orthogonal to the bounding circle C. (b) illustrates Axiom 4.4, the hyperbolic substitute for the parallels axiom.

However, the concept of *congruence* in Poincaré's model is not so simple. In preparation, we discuss *inversion in a circle*.

Inversion in a circle. In Figure 4.16, the point Q is the inverse of point P with respect to the circle C. Point P must be distinct from O, the center of circle C. Point Q is located on the extension of the radius through P such that the product of the two distances from the origin O to the points P and Q, respectively, is equal to the square of the radius of the circle C. In other words, we have

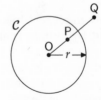

Figure 4.16. Inversion in a circle.

$$\overline{OP} \cdot \overline{OQ} = r^2$$

The relationship between P and Q is reciprocal. That is, P is also the inverse of Q with respect to C.

Inversion is a particular method of associating points in the plane with well-defined image points. In mathematics, such an association is called a *mapping* or a *transformation*. Under inversion, the image $\mathcal{I}$ of a figure $\mathcal{F}$ consists of the aggregate of the images under inversion of all the points of $\mathcal{F}$ (ignoring O, if that point happens to belong to $\mathcal{F}$). We say that inversion with respect to the circle C maps $\mathcal{F}$ into $\mathcal{I}$. Inversion with respect to a circle C has a number of remarkable properties.

Proposition 4.1 (properties of inversions). *Let C be the circumference of a circle with center at O (as in Figure 4.16).*

1. *Inversion with respect to the circle C maps each point of the circumference of C into itself and maps the interior of the circle C onto its exterior.*
2. *The inverse with respect to C of a line or circle is a line or a circle. More specifically:*
   (a) *The inverse with respect to the circle C of a circle $\mathcal{D}$ not intersecting O is a circle $\mathcal{E}$ not intersecting O. The interior of the circle $\mathcal{D}$ is mapped onto the exterior or interior of the circle $\mathcal{E}$ depending on whether O is in the interior or exterior of $\mathcal{D}$.*
   (b) *The inverse of a circle intersecting O is a line not intersecting O.*
   (c) *The inverse of a line not intersecting O is a circle intersecting O.*
   (d) *Inversion with respect to C maps each line intersecting O into itself.*
3. *Inversion preserves angles between straight or curved lines. For example, the inverse with respect to C of two orthogonal circles not intersecting O consists of two orthogonal circles not intersecting O.*

Congruence in Poincaré's model.   The simplest instance of congruence in Poincaré's model is realized by inversion with respect to a hyperbolic line. Recall hyperbolic lines are circular arcs orthogonal to the bounding circle C. This kind of inversion is like reflection across a line in the Euclidean plane. *In general, a hyperbolic congruence is the composition of finitely many inversions across one or more hyperbolic lines.*

Figure 4.17(a) shows a hyperbolic congruence of the simplest type — inversion with respect to a single hyperbolic line $\mathcal{H}$. The figures PQR and P′Q′R′ are hyperbolic triangles because all of the sides (e.g., PQ) are hyperbolic line segments, that is, arcs of circles orthogonal to C. Although they are clearly not congruent triangles in the Euclidean sense, these hyperbolic triangles are congruent because P′Q′R′ is the inverse of PQR with respect to the hyperbolic line $\mathcal{H}$.

The sum of the angles of a hyperbolic triangle is always *less* than 180°. Notice that corresponding angles of the two congruent triangles are equal.

In Figure 4.17(a), the hyperbolic line (circular arc) $\mathcal{H}$ divides the interior of C into two parts. Inversion with respect to $\mathcal{H}$ interchanges these two parts.

Figure 4.17(b) shows hyperbolic congruence in a more general sense. The three circles C_1, C_2, and C_3 are hyperbolically congruent, but, because they are not all the same size, they are not congruent in the Euclidean sense. In fact, C_2 is the inverse of C_1 with respect to the hyperbolic line $\mathcal{H}_1$, and C_3 is the inverse of C_2 with respect to $\mathcal{H}_2$. The circles C_1 and C_3 are congruent because a composition of two inversions maps the one onto the other. (To establish congruence, we do not need to address the question whether *one* inversion would be sufficient.)

On the right edge of Figure 4.17(b), between the smallest circle C_3 and

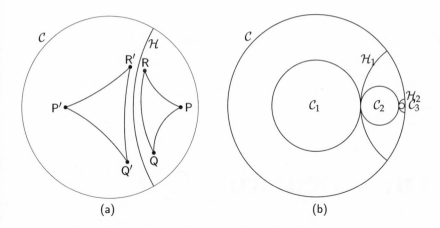

(a)   (b)

Figure 4.17. Congruent figures in Poincaré's model of hyperbolic geometry. (a) Triangles PQR and P′Q′R′ are congruent in the hyperbolic sense. The congruence is achieved by inversion with respect to the hyperbolic line $\mathcal{H}$. (b) The three circles $\mathcal{C}_1$, $\mathcal{C}_2$, and $\mathcal{C}_3$ are congruent in the hyperbolic sense. In particular, $\mathcal{C}_1$ and $\mathcal{C}_2$ are transformed, respectively, into $\mathcal{C}_2$ and $\mathcal{C}_3$ by inversions with respect to the hyperbolic lines $\mathcal{H}_1$ and $\mathcal{H}_2$, respectively.

the bounding circle $\mathcal{C}$, there is a gap, so small that is it barely visible. Figure 4.18 shows a magnification of this gap. In this gap, it is theoretically possible to inscribe a fourth circle $\mathcal{C}_4$ congruent to $\mathcal{C}_1$, and so on.

Hyperbolic distances become larger and larger relative to Euclidean distances the closer we approach the bounding circle $\mathcal{C}$. The bounding circle $\mathcal{C}$ is infinitely far (in the hyperbolic sense) from any point in the interior of $\mathcal{C}$. Like Euclidean lines, hyperbolic lines, such as $\mathcal{H}_1$ and $\mathcal{H}_2$ in Figure 4.17(b), have infinite extent.

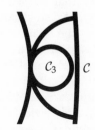

This chapter gave an introduction to curvature of curves and surfaces and to non-Euclidean geometry. We have also seen the meaning of intrinsic and extrinsic geometry. We can draw two important conclusions about the geometry of our universe:

Figure 4.18. Blowup of detail of Figure 4.17(b), showing gap between $\mathcal{C}_3$ and $\mathcal{C}$.

1. Only the intrinsic geometry of the universe is meaningful.
2. The geometry of Euclid does not hold a privileged place.

In the next chapter, we will see an aspect of geometry that is basic to science, yet of very recent origin: the usage of graphs.

5

The Calculating Eye

One picture is worth ten thousand words.

—FREDERICK R.BARNARD, Printers Ink, 10 March 1927

IN this chapter, we look at a mathematical innovation that promotes visual–spatial thinking—a geometric innovation that is used primarily for purposes that are not geometric and not even mathematical—an innovation that is so much a part of modern life that it is difficult to think that there was a time when it did not exist. I refer to the commonplace graphs that we see in every newspaper—for example, graphs showing the price of stocks and bonds day by day or hour by hour. A graph expresses the relationship between two variables with an effectiveness that is impossible to achieve with words alone. Graphs give instant insight into complex matters. Surprisingly, graphs were seldom seen before the twentieth century.

Figures 5.1 and 5.2 are examples of a common type of graph with horizontal and vertical calibrated axes. A point on the curve indicates a correspondence between the values of the numerical variables x and y.[1]

This chapter also examines *coordinate geometry*, which extends classical geometry through application of the graph concept. We see this only in hindsight because, historically, coordinate geometry *preceded* the use of graphs.

A well-conceived graph demonstrates "One picture is worth ten thousand words."[2] The obvious meaning of this saying is that a picture makes clear what words cannot. However, a computer scientist might contend, on the contrary, that words communicate more efficiently than pictures—that pictures are a wasteful use of computer resources because to encode a picture electronically can actually require more bytes than do 10,000 words. To make these matters more specific, five letters per word is a

82

reasonable estimate for an average English sentence—six characters including spaces to separate words. Furthermore, the usual computer coding uses one *byte* of storage for each letter. A good quality image of Botticelli's *The Birth of Venus* can be encoded with 217,000 bytes. Dividing by 6 (the number of bytes in an average word), we find that *The Birth of Venus* is equivalent to about 36,000 words. For comparison, Mark Twain's *Huckleberry Finn* uses 570,000 bytes—about 95,000 words. By this reckoning, if one is collecting classic paintings and novels on a computer, *Huckleberry Finn* fills about 2.6 times as much disk space as *The Birth of Venus*. But two or three pictures could not possibly tell a complex story like *Huckleberry Finn*—certainly more than 100 pictures would be needed.

The electronic costliness of pictures compared with text underlies the fact that early computer displays were *text only*. Only when electronic storage devices became cheaper and more efficient was it feasible to display information on a computer like the Apple Macintosh using a *graphical user interface*—in computer jargon, a *GUI*.

Of course, the above analysis is simplistic. We do not measure the nourishment of food by the number of bites, and if we judged paintings and novels by the number of computer bytes, then art and literary critics would have nothing to do. Even though pictures are a more costly use of computer resources, we cannot ignore the fact that pictures and words convey *qualitatively* different information. In fact, medical science has shown that pictures and words are processed by different parts of the brain.[3]

Visual–spatial thinking dominates mathematical thinking at the highest level. Recent medical research (Witelson, Kigar, and Harvey (1999)) reports that the preserved brain of Albert Einstein shows unusual development in the parietal lobes—known to be the site of visual–spatial cognition.[4] This is consistent with the following statement of Einstein concerning his creative process:

> The words or the language, as they are written or spoken, do not seem to play any role in my mechanism of thought. The psychical entities which seem to serve as elements in thought are certain signs and more or less clear images which can be "voluntarily" reproduced and combined.

Einstein obtained his ideas from a visual source in his mind, but he needed to communicate these ideas in words—with great difficulty, so he tells us. Difficult to explain and difficult to understand. One used to hear, "There are only *n* people in the world who understand Einstein's theories," where *n* is some number less than 20. Or, to put it more irreverently,

There's a wonderful family named Stein,
There's Ep, there's Gert, and there's Ein.
  Ep's statues are junk,
  Gert's poems are bunk,
And nobody understands Ein.[5]

Since visual thought lies behind his mathematical discoveries, perhaps Einstein could have made his ideas less difficult if he had used more pictures — if the mathematical journals would have let him. Although mathematical journals value a high level of creativity, published articles contain much step-by-step verification of correctness — with few clues of the author's discovery process. Journals discourage the use of pictures and generally force authors into an unattractive succinct style for two reasons:

1. To enforce the traditional impersonal style. A scientific article should be judged on its scientific merit. The personality of the author must not intrude.
2. To reduce the expense of typesetting.

But reason 1 should not force the exclusion of pictures that aid understanding. Moreover, reason 2 is less valid today than formerly because now authors can typeset their own work with powerful computer programs that include a rich set of tools for typesetting diagrams and mathematical formulas. For example, this book was typeset by the author using LATEX and MetaPost.

Einstein's reflection on his source of inspiration is a glimpse of a little-known truth — that it takes more than logical thinking to solve mathematical problems.[6] Music is more than notes written on a staff, and mathematics is more than formal proofs and algebra. This is true for the discovery of the theory of relativity, but it is also true for the solution of problems in elementary mathematics. Mathematical excellence requires hard work and careful study; nevertheless, the elusive spark of insight often comes from a free-floating visual–spatial intuition.

Graphs

Graphs are a surprisingly recent invention. Graphs did not exist in antiquity. Although the ancient Greek mathematicians used complex diagrams abundantly, they did not use graphs — at least, not explicitly. Nevertheless, a graph is implicit, for example, in the method attributed to Archytas (428?–347? BCE),[7] illustrated in Figure 5.1, for finding the geometric mean of two quantities.

Archytas's solution can be expressed very naturally using a graph. By definition, the geometric mean of two positive numbers is equal to the square root of their product. In Figure 5.1, y is the geometric mean of x

and $1 - x$, that is, $y = \sqrt{x(1 - x)}$. The result of Archytas is equivalent to the assertion that the graph in Figure 5.1 is a semicircle.[8]

The graph in Figure 5.1 would not have made sense to Archytas and his contemporaries, and not even to mathematicians that came two thousand years later. Figure 5.1 uses concepts that came together only at the end of the seventeenth century.

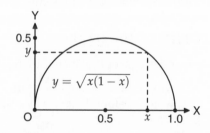

Figure 5.1. The semicircle with radius $1/2$ and center at $(1/2, 0)$ is the graph of the geometric mean of x and $1 - x$.

We can compare the innovation of graphs with other human inventions—for example, with the prehistoric invention of *basket making*. There were three essential elements in this discovery.

1. There was a *need*. Baskets were useful, for example, for gathering and preparing food.
2. *Materials* for making baskets were available.
3. There were *clever people* who developed the technique for making baskets.

The need for graphs

Graphs would have benefited Galileo, but they were not available to him. He made empirical measurements of balls rolling down a ramp. Part of his great genius was that he saw the need for this experiment. He had no reason to expect a simple underlying principle. Yet he was able to find the principle, even with primitive intellectual tools. Graphs appeared about a century later. Consequently, Galileo did *not* illustrate his result by means of a graph such as Figure 5.2.

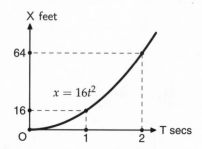

Figure 5.2. Graph showing the downward motion of a falling object.

The need for graphs depended on the state of science and mathematics together with the attitudes and endeavors of scientists and mathematicians.

- *Empirical scientists need graphs.* Their measurements can appear as dots on a graph. By somehow connecting the dots, the researcher can obtain a global view of the phenomenon being studied and can be led to general underlying principles.

- *Mathematicians need graphs.* For example, an equation takes on a concrete meaning when we see its graph, and consequences of the equation become more apparent.

"Materials" for graphs

Graphs need conceptual rather than physical materials:

1. A well-tended menagerie of mathematical *curves.*
2. A well-developed *number system.*
3. *Algebraic calculation.*
4. An interest in *empirical measurement.*

Curves

The study of curves began in antiquity. In the ancient Greek mind, mathematical diagrams (lines, circles, etc.) depicted only a small number of objects with abstract mathematical definitions. The idea of an *arbitrary* relationship between two variables or an arbitrary curve did not gain mathematical legitimacy until the middle of the nineteenth century.

For the ancient Greeks, there were only a small number of mathematical curves. The ancient Greeks studied not only lines and circles but also ellipses, parabolas, and hyperbolas, curves that we will consider further on pages 101–107. Although these curves received the greatest attention, the Greeks also considered a few other curves, for example, the Spiral of Archimedes (Figure 5.3(a)).

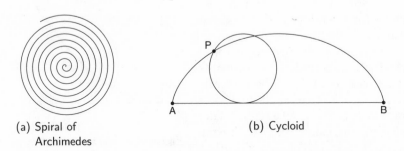

(a) Spiral of   (b) Cycloid
    Archimedes

Figure 5.3. Mechanically generated curves.

(a) The spiral of Archimedes is generated by a point on a disk rotating with uniform angular velocity. Relative to the moving disk, the tracing point moves with uniform linear velocity along a line from the center of the disk to a fixed point on the circumference of the disk. The groove on a phonograph record is approximately an Archimedean spiral.

(b) The point P traces the cycloid AB as the circle rolls without slipping from point A to point B. A point on the surface of a rolling automobile tire traces a cycloid.

Mathematicians of the seventeenth century added a number of curves to the list — especially mechanically generated curves like the cycloid (Figure 5.3(b)), the path traced by a point on the circumference of a rolling circle. The name *cycloid* was coined by Galileo in 1599. Galileo attempted to find the area of the cycloid, even resorting to weighing models cut in the shape of a cycloid, but failed to discover the fact that this area is exactly three times the area of the rolling circle. Before the seventeenth century, legitimate mathematical curves had either geometric or mechanical definitions. Curves generated by algebraic equations were slow to be included in the canon — even Newton rejected them. Seventeenth-century mathematicians treated curves as mathematical playthings with curious properties, but they were slow to see curves as valuable tools for understanding matters beyond the curves themselves.

The number system

Graphs need a well-developed number system that serves as model for points on a line. A suitable number system enables the definition of *coordinates* of points in a plane. Coordinates are pairs of numbers that specify points in the plane. The modern concept of *real numbers* is ideally suited to this identification between pairs of numbers and points in the plane. Lacking the modern concept of real numbers, a clear idea of decimal fractions would suffice, a notion that goes back to antiquity. Although decimal fractions and algebra were available in Galileo's time, he did not make use of them; he recorded his observations of balls rolling down a ramp using only whole numbers.

Graphs also require an understanding of *negative numbers*. Mathematical pioneers of the seventeenth century, such as Descartes and Fermat, still lacked an understanding of negative numbers.

Algebraic calculations

In the distant past, there was not agreement among mathematicians concerning what sort of calculations should be given mathematical legitimacy. The ancient Greeks gave legitimacy to geometric calculations and to the arithmetic of the rational numbers. In the ninth century, the Arab mathematician al-Khwarizmi (780?–850?) wrote about algebra, but he justified his calculations geometrically; algebraic calculations were not yet considered legitimate mathematics.

Descartes developed an algebraic notation that is fully recognizable to today's algebra student. Descartes understood that algebra provides a way of studying geometrically or mechanically defined mathematical curves, but he missed the idea that algebra provides a means of defining a vast universe of new curves and much more — a powerful new tool for understanding the world.

In Europe in the seventeenth century, algebraic calculations were still controversial; this issue was the basis, at least in part, of the quarrel between mathematician John Wallis (1616–1703), who supported algebra, and philosopher Thomas Hobbes (1588–1679), who opposed it. By the beginning of the eighteenth century, all mathematicians considered algebraic calculations legitimate.

Empirical measurement

Mathematics often uses graphs to depict an equation, but in every newspaper graphs are used in a more commonplace manner — to display empirical information.

Platonic idealism elevated geometry and denigrated empiricism. For the Greeks, an important obstacle to empirical science was the attitude, expressed by Plato, that philosophy and mathematics had little or no need of the lowly craft of empirical observation. It is undoubtedly true that the advances in empirical science in the twentieth century have been accompanied by a blossoming in the use of graphical representations.

Cartography is an ancient science that joined empirical measurement with geometry. The ancient Greeks had both geometric drawings and cartography. In fact, both of these precursors of graphs came together in the person of Claudius Ptolemy (100?–170? CE). Ptolemy was both an astronomer–mathematician and a geographer. He was the author of *Almagest*, 13 books detailing his geocentric theory of the solar system, and *Geography*, eight books containing a collection of cartographic data and maps. Ptolemy's astronomy was invalidated by Copernicus, but we still use his cartographic concepts of latitude and longitude. It seems that the ancients believed that curves in maps had nothing whatever to do with the curves of mathematics.

One may ask, what prevented Ptolemy, who furthered the science of cartography, from discovering graphs? Since many graphs, like Figure 5.2, express the growth of something over time, the ancients were hampered because they lacked an accurate means of measuring time. Mechanical clocks were not invented until the late Middle Ages. Galileo measured short intervals of time by observing the amount of water flowing out of a tank. On page 177, we will see that he also used his sense of rhythm to determine the equality of short time intervals.

Mathematical diagrams and maps both resemble the object under discussion. A circular drawing approximates the Platonic ideal circle, and a map of Egypt approximates the shape of the land that it portrays. The graph concept requires a further leap of abstraction because a graph is not a visual likeness.

Clever people invented graphs

The representation of data using graphs was invented in the Middle Ages by Nicole Oresme (1320?–82), Bishop of Lisieux in Normandy, but the idea remained unexploited for several centuries, probably because in the middle ages no one could foresee the immense potential of this discovery.

The first scientist to make extensive use graphs was the German mathematician Johann Heinrich Lambert (1728–77). (See Figure 5.4.) The Scottish political economist William Playfair[9] (1759–1823) was the first to publish statistical graphs in 1785. An example of his work is shown in Figure 5.5. Florence Nightingale (1820–1910) is renowned as a reformer of hospitals and nursing, but it is not as well known that she was a pioneer in the graphical presentation of medical statistics. She used a fan-shaped graph that she called a "coxcomb" (Figure 5.6) to document the casualties of the Crimean war (1853–56).

According to the *Oxford English Dictionary* (OED), the term *graph of a function* was first used by mathematician George Chrystal (1851–1910) in 1886. Chrystal's usage of the word *graph* is the familiar one that we use in this chapter. The verb *to graph* first appeared in 1898.

According to the OED, the term *graph* was used even earlier by math-

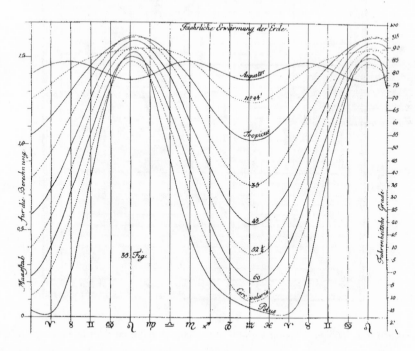

Figure 5.4. From Lambert (1779). Solar warming throughout the year at different latitudes.

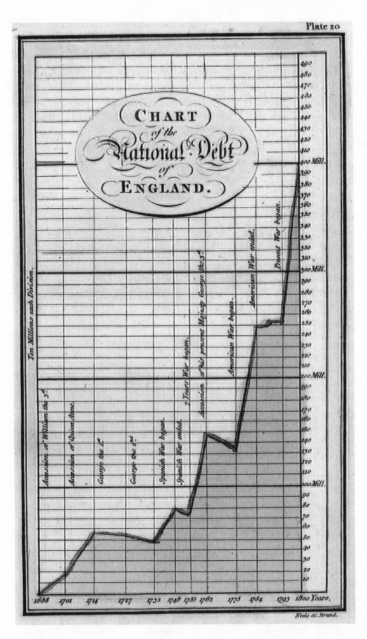

Figure 5.5. William Playfair's graph of the British national debt in Playfair (1801).

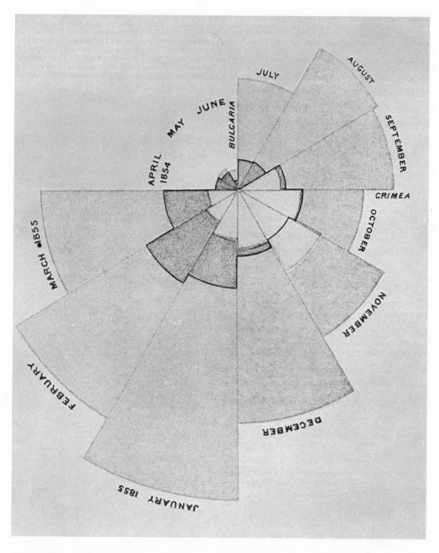

Figure 5.6. Florence Nightingale's "coxcomb" graph showing the casualties of the Crimean war month by month. The outer, inner, and middle areas represent, respectively, mortality due to disease, wounds, and other causes. Nightingale (1858).

ematician James J. Sylvester (1814–97) in 1878, but not in the sense used in this chapter. In this OED citation, Sylvester uses *graph* for a chemical diagram that is now called a *structural formula*. Sylvester's usage presaged the more recent mathematical meaning of the word (Kőnig (1950)). In this sense, a graph consists of a (usually finite) collection $\mathcal{P}$ of points together with a collection of line segments connecting pairs of points in the set $\mathcal{P}$.

Also in the late nineteenth century, the practice originated of drawing graphs using paper ruled with squares — used much earlier in the arts and crafts. Indeed, such paper is now called *graph paper*. For today's students, graphing problems are a standard homework assignment, but such figures were seldom seen in print before 1900. The absence of graphs holds not only for mathematics but also science generally and even for the financial pages of newspapers. The *New York Times* started printing graphs in its financial section only in the early 1930s.[10] Figure 5.7 shows the increase through the years 1879–1957 of the incidence of graphs in the renowned

		Pages in *Nature*		
Volume	Year	total	with graphs	percent
20	1879	644	1	0.2
40	1889	660	2	0.3
60	1899	699	14	2.0
80	1909	560	2	0.4
100	1917	520	5	1.0
120	1927	976	20	2.0
140	1937	1112	44	4.0
160	1947	916	107	11.7
180	1957	1498	267	17.8

(a)

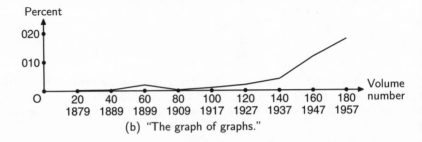

(b) "The graph of graphs."

Figure 5.7. Incidence of graphs in the science journal *Nature*. The table shows a count of pages with at least one graph starting with Volume 20 in 1879.

British journal of science *Nature*. The incidence of graphs increased from almost none in 1879 to near the current level in 1957.

Why have scientists and mathematicians been so slow to see the benefit of graphs? There are several possible reasons:

1. Printing costs are lower for text than for diagrams.

2. The abstract mathematical concept of *function*—a correspondence between an independent and a dependent variable—made its rigorous debut in the latter half of the nineteenth century. The concept of *function* gives mathematical legitimacy to graphs.

3. In the twentieth century, science and technology have produced a flood of data that did not exist in earlier times. The volume of data grew because technology made it possible, and technology grew in part because analysis of the data led to further advances. It became urgent to find the most efficient methods of understanding data and drawing conclusions from it. Highly sophisticated methods of statistical analysis grew out of this need, but also the simpler and older ideas of graphical presentation helped scientists and others to make sense of the proliferation of data.

Coordinate Geometry

> *But I shall not stop to explain this in more detail, because I should deprive you of the pleasure of mastering it yourself, as well as the advantage of training your mind by working over it, which is in my opinion the principal benefit to be derived from this science.*
>
> —RENÉ DESCARTES, La Géométrie, 1637

Coordinate geometry introduced the familiar orthogonal coordinate system previously seen, for example, in Figure 5.2. Figure 5.8 shows a point with coordinates (x, y) where x and y are the distances of the point from the coordinate axes.[11] The use of coordinates is so universal now, it is difficult for us to appreciate the depth of this innovation. It is a technique unknown to the ancient Greek mathematicians.

Figure 5.8.
Coordinate axes.

Coordinate geometry, also known as *analytic* geometry, was developed simultaneously and independently by René Descartes (1596–1650) and Pierre de Fermat (1601–65).[12] They observed that geometric objects (lines, circles, etc.) can be described by *algebraic equations* and that geometric theorems can be proved algebraically—a brilliant new technique that eluded the ancient Greek geometers.

Nevertheless, both Descartes and Fermat would find much to learn today in an elementary class in analytic geometry. Their first lack would be the concept of *graph*. They believed that curves defined by geometric or mechanical means could be studied by algebraic means, but they did not use algebra as a means of generating new curves, and they did not consider curves as a tool for understanding algebra or science. For example, a graph like Figure 5.2 would depict an unfamiliar new idea. Furthermore, they would be greatly hampered by their unfamiliarity with negative numbers. In their first few class sessions, Descartes and Fermat would see for the first time the formulas for the distance between two points and for the angle between two lines. They would be unfamiliar with the rectangular coordinate system shown in Figure 5.8. They would even be unfamiliar with the identification of points in the plane with pairs (x, y) of numbers, a concept lying at the very foundation of analytic geometry. Considering all this, how can Descartes and Fermat be the discoverers of analytic geometry? The answer is that they solved a number of geometric problems by algebraic means. They used coordinates of points, but only *implicitly*.

Synthetic versus analytic

Although *coordinate* geometry is more descriptive, the term *analytic geometry* is more usual. In contrast, the geometry introduced by the ancient Greeks is called *synthetic*. All of the proofs in Euclid's *Elements* are synthetic. Initially, coordinate geometry was handicapped by the lack of an ancient tradition.

Etymology

The mathematical meanings of *synthetic* and *analytic* are not the same as the general meanings:

> **synthesis** — assembling constituent parts into a whole.
> **analysis** — separating a whole into constituent parts.

The mathematical meaning of the word *analysis* changed during the twentieth century. The 1933 edition of the *Oxford English Dictionary* has the following entry:

> **Analysis ... Math.**
> **Ancient Analysis,** *The proving of a proposition by resolving it into simpler propositions already proved.*
> **Modern Analysis,** *The resolving of problems by reducing them to equations.*

These definitions reflect the usage by mathematicians when this edition of the *OED* was compiled from 1879 to 1928. The term, *ancient analysis*, now obsolete, applied to the methods of Euclidean geometry, for example, in Euclid's *Elements* — methods that are now called *synthetic*. Today *modern analysis* refers primarily to the book of that name by Whittaker and Watson, first published in 1902; the content of that book is now called *classical analysis* — a topic that used to be considered much more central to mathematics than it is now. Prior to 1800, *analysis* was another word for *algebra*, but today the term *mathematical analysis* denotes calculus together with its advanced ramifications — today, analysis and algebra are considered two separate mathematical disciplines.

Synthetic and analytic proofs

We will contrast the synthetic and analytic proofs for a certain remarkable theorem — but first, two definitions:

Definition 5.1. Median. The line segment connecting the vertex of a triangle with the midpoint of the opposite side is called a *median* of the triangle.

For example, in Figure 5.9 the points P, Q, and R are the midpoints of the three sides of the triangle ABC. The line segment AP is one of the three medians.

Definition 5.2. Concurrent. Three or more lines are concurrent if they intersect in a single point.

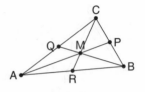

Figure 5.9.

In Figure 5.9, the medians are concurrent in point M, but this concurrency requires proof. We shall see first a synthetic and then an analytic proof of the following theorem:

Theorem 5.1. *The medians of a triangle are concurrent. The point of concurrency divides each median in the ratio 2:1. More specifically, for each median, the distance from the point of concurrency to the corresponding vertex is twice the distance from the point of concurrency to the corresponding midpoint.*

For example, in Figure 5.9, the length $\overline{AM}$ is equal to $2\overline{MP}$.

A synthetic proof is based on the Euclidean axioms and previously proved results. In particular, this proof makes use of the following two propositions:

Proposition 5.1. *A line segment bounded by midpoints of two sides of a triangle is parallel to the remaining side and half its length.*

For example, in Figure 5.10, if points Q and P are midpoints of AC and BC, respectively, then the segment QP is parallel to side AB and we have $\overline{AB} = 2\overline{QP}$.

Proposition 5.2. *The diagonals of a parallelogram bisect each other.*

Synthetic proof of Theorem 5.1. Referring to Figure 5.11 and using Proposition 5.1, the dashed line segment QP connecting the midpoints of the triangle is parallel to the side AB and half its length. Suppose that the medians AP and BQ intersect at point M and let S and T be the midpoints of AM and BM, respectively.

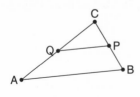

Figure 5.10.

Applying Proposition 5.1 again, we see that the dashed line segment ST is also parallel to the side AB and half its length. We see that QPTS is a parallelogram because sides QP and ST are parallel and have the same length. By Proposition 5.2, the lengths $\overline{SM}$ and $\overline{MP}$ are equal. Furthermore, $\overline{AS} = \overline{SM}$ because S is the midpoint of AM. In summary, we have

$$\overline{AS} = \overline{SM} = \overline{MP}$$

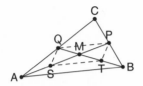

Figure 5.11.

It follows, as claimed, that M divides the median AP in the ratio 2:1.

Similarly, M divides the median BQ in the ratio 2:1. Furthermore, the third median CR (shown in Figure 5.9) intersects the other two in point M and is divided by M in the ratio 2:1. □

This proof is beautifully conceived, but the discovery process is hidden. One is tempted to say, "I could never have discovered this proof."

Now we turn to an analytic proof—a proof using coordinate geometry that the ancient Greeks would never have found. I think that it is again a beautiful proof, but shorter than the synthetic proof. Furthermore, the discovery process is more transparent. The analytic method involves a straightforward algebraic computation. It is more possible that one might say, "I could have done that."

The analytic proof makes use of the following proposition:

Proposition 5.3. *If three parallel lines intersect two transversals, then they divide the transversals proportionally.*

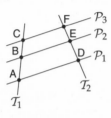

Figure 5.12.

For example, in Figure 5.12, $\mathcal{P}_1$, $\mathcal{P}_2$, and $\mathcal{P}_3$ are parallel lines intersected by the transversals $\mathcal{T}_1$ and $\mathcal{T}_2$. Proposition 5.3 asserts that the following proportion holds:

$$\overline{AB} : \overline{BC} :: \overline{DE} : \overline{EF}$$

Analytic proof of Theorem 5.1—first method. Referring to Figure 5.13, in triangle ABC, let

P be the midpoint of BC, and let M be the point that divides the median AP in the ratio 2:1, that is, $\overline{AM} = 2\overline{MP}$. Further, let M' and M'' (not shown in Figure 5.13) denote the points that divide the other two medians in the ratio 2:1. We do not know *a priori* that M, M', and M'' are different names for the same point; it is our task to show that all three are identical. Suppose that the x-coordinates of A, B, and C are, respectively, x_1, x_2, and x_3. Now we compute the x-coordinates of P and M, which we denote x_4 and x_5, respectively. Since P is midpoint of BC, using Proposition 5.3, we have $\overline{TU} = \overline{UV}$, which implies

$$x_4 = \frac{x_2 + x_3}{2} \tag{5.1}$$

Similarly, using the fact that $\overline{AM} = \frac{2}{3}\overline{AP}$, we have $\overline{RS} = \frac{2}{3}\overline{RV}$; or, in other words,

$$x_5 - x_1 = \frac{2}{3}(x_4 - x_1) \tag{5.2}$$

Substituting equation (5.1) into equation (5.2), we have

$$x_5 = x_1 + \frac{2}{3}\left(\frac{x_2 + x_3}{2} - x_1\right) = \frac{x_1 + x_2 + x_3}{3} \tag{5.3}$$

Similarly, if we were to compute the x-component of the points M' and M'', we would find the same value $\frac{1}{3}(x_1 + x_2 + x_3)$. This is true because of the *symmetry* of the formula $\frac{1}{3}(x_1 + x_2 + x_3)$. Therefore, the three points, M, M', and M'', have the same x-coordinate.

A similar computation shows the y-components of M, M', and M'' are all equal to $\frac{1}{3}(y_1 + y_2 + y_3)$. Since the three points M, M', and M'' have the same coordinates, they must be identical.   □

The above analytic proof shows clearly that the concept of symmetry is central—symmetry not of the figure but of an algebraic formula. The proof

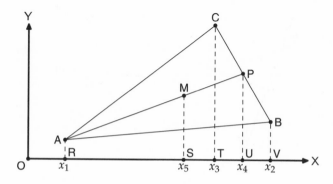

Figure 5.13.

is undoubtedly logically correct. However, it could be faulted, because it conceals one aspect of the discovery process: Where did the number 2/3 come from? We can remedy this with further use of the analytic method. The following proof *discovers* the number 2/3 in a natural way instead of pulling it out of a hat. In the preceding proof, the coordinate system had no special relationship with the triangle. However, the cleverness of the following proof consists in choosing the coordinate system in a special way.

Analytic proof of Theorem 5.1—second method. In Figure 5.14, points P and Q are midpoints of sides BC and AC, respectively. Departing from the preceding proof, the point M is defined to be the intersection of the medians AP and BQ. (This is a departure from the preceding proof in which M was defined as the point that divides the median AP in the ratio 2:1.)

We choose the coordinate system so that vertex A is the origin and the y-axis is in the direction of the median AP. As before, we denote the x-coordinates of the points A, B, C, P, and M, as $x_1, x_2, x_3, x_4,$ and x_5, respectively. In addition, x_6 denotes the x-coordinate of Q. Note that we have

$$x_1 = x_4 = x_5 = 0$$

Since P is the midpoint of BC, from Proposition 5.3, we have $\overline{AT} = \overline{RA}$, and it follows that $x_2 = -x_1$. Again, since Q is the midpoint of AC, we have $\overline{RS} = \overline{SA}$, and it follows that $x_6 = \frac{1}{2}x_3$. Therefore, $\overline{AT}$ is equal $2\overline{SA}$, and it follows, again from Proposition 5.3, that $\overline{MB}$ is equal $2\overline{QM}$.

We have shown that the intersection of the *median* AP with the median BQ divides BQ in the ratio 2:1. But the choice of these particular medians

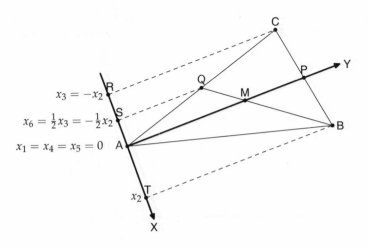

Figure 5.14.

was arbitrary. It follows that the intersection of two medians divides each in the ratio 2:1, and it follows that all three medians are concurrent.   □

Straight lines

Straight lines are an abstraction of various objects of experience — the shape of a taut string, the folded edge of a flat piece of paper, or dust motes in a ray of sunlight. The uses of straight lines by the carpenter, the surveyor, and the draftsman were important reasons for the development of Euclidean geometry in antiquity. However, there is an even more important presence of straight lines in ordinary life, familiar to us (but not to the ancient Greeks): motion of constant velocity.

In particular, the graph of a motion of constant velocity, illustrated in Figure 5.15, is an example of a straight line that is just as natural as the examples listed above. In the figure, the initial distance is s_0, and the initial time is t_0. The motion has constant velocity if the distance traveled is proportional to the elapsed time — that is, if the ratio of the distance $s - s_0$ to the time $t - t_0$ is a constant, which we will call v. The requirement that v is constant gives us the *equation* of the straight line $\mathcal{L}$:

$$s = s_0 + v(t - t_0) \qquad (5.4)$$

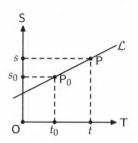

Figure 5.15. Motion of a particle with constant velocity. The distance $s - s_0$ traveled in time $t - t_0$ is a constant v independent of the choice of the points P_0 and P.

This equation expresses a relationship between the variables s and t. The initial position s_0, the initial time t_0, and the velocity v are constants. The velocity can be positive or negative depending on whether s increases or decreases over time. If s decreases as time increases, then the velocity v is negative. The magnitude of the velocity v, ignoring its positive or negative sign, is called the *speed* of the motion.

Figure 5.15 specifies that the variable t is time and s is distance. However, there are many other ways to interpret s and t. For example, s and t could both represent distances. This is the most straightforward interpretation because in the paper and ink drawing s and t represent actual physical distances. If s and t are distances, then v is called the *slope* of the line. (However, it is customary to use the letter m to represent slope.)

On a highway, if s represents vertical climb corresponding to horizontal distance t, then v is called the *grade* and is usually expressed in percent. For example, a 10% grade signifies 1 foot elevation gain for every 10 feet of horizontal progress. In general, whatever the variables s and t signify, v is the *rate* of s with respect to t. In Figure 5.15, this rate is constant. In later chapters, we will see how calculus deals with rates that are not constant.

The trains and the whiffle bird

The following example makes use of the above linear representation of constant velocity (see Figure 5.15) and shows the benefit of representing a problem graphically.

Question 5.1. Two trains, the *Super Chief* (*SC*) and the *Twentieth Century Limited* (*TCL*), are 50 miles apart and are moving toward each other on the same track at velocities of 20 and 30 miles per hour, respectively. A whiffle bird starting at the front of the *Super Chief* flies back and forth between the trains at a rate of 200 miles per hour until the trains collide. What is the total distance the bird has flown?

There is an easy way and a hard way to solve this question. The hard way is to compute the time to complete each flight — there are infinitely many of them — and then to sum this infinite series, a lengthy but feasible calculation.

The easy way is to observe that the trains are approaching each other with a combined velocity of $20 + 30 = 50$ miles per hour. Therefore, they will traverse the 50 miles separating them and collide in exactly one hour, in which time the whiffle bird, at 200 miles per hour, travels 200 miles.

This question relates to a story concerning the legendary calculating powers of a certain mathematician. A distinguished physicist observed that physicists generally solve the question the easy way and mathematicians solve it the hard way. He posed the above problem to this mathematician who immediately responded, "200 miles."

"This is strange," said the poser, "because mathematicians generally sum the infinite series."

"What do you mean, strange?" replied the mathematician. "That's how I did it!"

One is more likely to find the easy solution on seeing the graphical representation of the problem shown in Figure 5.16, which shows the positions of the two trains and the bird at each moment of time. The two straight lines and the zigzag line represent the motions of the two trains and the bird, respectively.

The following mathematical argument shows that the remarkable whiffle bird possesses extraordinary powers.

Proposition 5.4. *The whiffle bird makes infinitely many flights.*

Proof. Suppose that there exists a final flight. We will show that this assumption leads to a contradiction. On its last flight the bird must leave

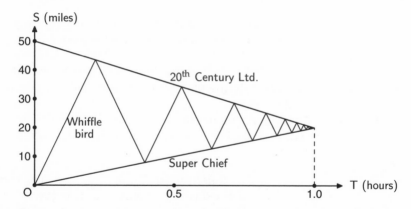

Figure 5.16. The collision of the *Super Chief*, the *Twentieth Century Limited*, and the whiffle bird.

one train and head toward the other. Specifically, suppose that on its last flight the bird leaves the *SC* and heads toward the *TCL*. The bird's velocity is greater than the velocity of the *SC*, and, therefore, the bird must meet the *TCL* before the *SC* does. In other words, the bird meets the *TCL before* the collision. It follows that the bird reverses its direction and makes still another flight contrary to the assumption that the current flight is the last. This concludes the proof by contradiction.   □

Conic sections

Clocks and plates — our world abounds with circular objects. But in a photograph or a drawing, a circle is generally represented as an *ellipse*, the perspective representation of a circle. In fact, the eye *sees* a circle as an ellipse. We acknowledge the skill of a still-life painter when he draws a fruit bowl as an almost perfect ellipse. The ellipse is a precisely defined curve, not simply an elongated circle. The ellipse belongs to a family of curves called the *conic sections*, or, more briefly, the *conics*. We will also examine the other members of this family — the *parabola* and the *hyperbola*.

Coordinate geometry is the most powerful tool for establishing the properties of the conics. A large part of a standard course in analytic geometry is devoted to these curves. We will see how the conics can be represented using coordinate geometry, but first we show what these curves have to do with cones using the same methods used by Apollonius of Perga (260?–185? BCE) and the other ancient Greek geometers who first investigated the conics. The conic sections are so named because these curves can be obtained as the intersection of a plane and a circular cone.

In the following discussion, our usage of the word *cone* differs from the ordinary meaning:

- In the following discussion, a cone consists of *two* parts, called *nappes*, as shown in Figure 5.17. In other contexts, a cone often has just *one* nappe.
- The two circular disks, the top and bottom shown Figure 5.17, are *not* part of the cone. In the present discussion, the cone consists of the lateral surface *only*.
- The cone considered here has infinite extent. Figure 5.17 shows just part of the entire cone. The cone continues indefinitely upward and downward.

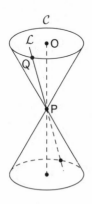

Figure 5.17. A cone of two nappes.

Referring to Figure 5.17, a circular cone is defined with reference to a circle C with center O and a second point P, the vertex of the cone. The point P lies on the line, called the *axis* of the cone, through O perpendicular to the plane of the circle C. The cone is the set of points swept out by the totality of lines intersecting the circle[13] C and containing the vertex P. Such a line is called an *element* of the cone. For example, line $\mathcal{L}$ is an element of the cone in Figure 5.17 because $\mathcal{L}$ contains the vertex P and intersects the circle C at point Q. All elements of a cone intersect the plane of the circle C at the same angle.

Different types of conic sections are obtained by intersecting the cone with planes having different orientations. Specifically, if the plane passes through the vertex P we obtain the three degenerate cases: (a) the vertex P only, (b) a single line element, or (c) two line elements—depending, respectively, on whether the plane is (a) less steep, (b) exactly as steep, or (c) more steep than the elements of the cone. In the following sections, we will see that if the intersecting plane does not contain the vertex P, then the three conditions of steepness mentioned above characterize, respectively, (a) the ellipse, (b) the parabola, and (c) the hyperbola.

The ellipse

An ellipse is obtained, as in Figure 5.18, by intersecting a cone with a plane $\mathcal{P}$ less steep than the elements of the cone. (All the elements of the cone are equally steep.) The plane does not contain the vertex of the cone, and since the plane is less steep than the elements of the cone, it intersects only one nappe of the cone. A circle is a special case of an ellipse generated when the plane $\mathcal{P}$ is perpendicular to the axis of the cone.

The ellipse has twice played a crucial role in the history of science:

1. The German astronomer Johannes Kepler (1571–1630) found by empirical observation that the orbits of the planets are ellipses. Isaac Newton (1642–1727) sought an explanation for Kepler's discovery. In doing so, he found the *law of universal gravitation*, which says that there is a force of attraction between any two masses directly proportional to the product of their masses and inversely proportional to the square of the distance between them. More specifically, there is a universal constant G such that masses m_1 and m_2 separated by the distance R are subject to a force of attraction equal to

$$\frac{m_1 m_2}{R^2}$$

2. Later, astronomers discovered that the planets, especially the planet Mercury, depart slightly from the predicted elliptical orbits. This minute difference could not be explained using Newton's law of gravitation. Seeking to explain this discrepancy, in 1916 Albert Einstein conceived of the *general theory of relativity*.

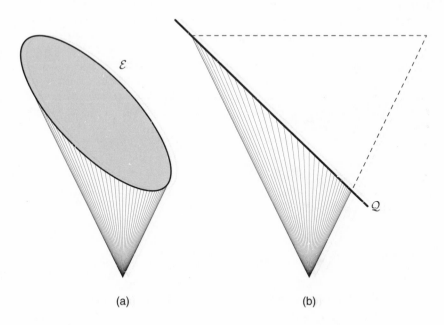

(a)   (b)

Figure 5.18. The ellipse $\mathcal{E}$ is the intersection of a cone with a plane $\mathcal{Q}$ less steep than the elements of the cone. The figure (a) is rotated by 45° in (b) showing the plane $\mathcal{Q}$ "on edge." The dashed lines in (b) indicate the portion of the cone cut off by the plane.

The parabola

A parabola is obtained, as in Figure 5.19, by intersecting a cone with a plane $\mathcal{P}$ parallel to an element of the cone. The plane does not contain the vertex of the cone and intersects only one nappe of the cone.

Ignoring the effect of air resistance, the trajectory of a thrown ball is a parabola.

The hyperbola

A hyperbola is obtained, as in Figure 5.20, by intersecting a cone with a plane $\mathcal{P}$. The plane does not contain the vertex of the cone, and since the plane is steeper than the elements of the cone, it intersects both nappes of the cone, creating two disconnected branches of the hyperbola. At points far from the vertex, the hyperbola approaches closer and closer to two intersecting straight lines called *asymptotes*.

Some comets have hyperbolic or parabolic orbits. These comets do not remain a part of the solar system. They approach from and return to deep space, making one pass around the sun.

The conic sections can be easily demonstrated with an ordinary electric lamp. A circular lampshade creates a cone of light. The shadow of the lampshade on a wall is elliptic, parabolic, or hyperbolic, depending on the orientation of the wall relative to the lamp.

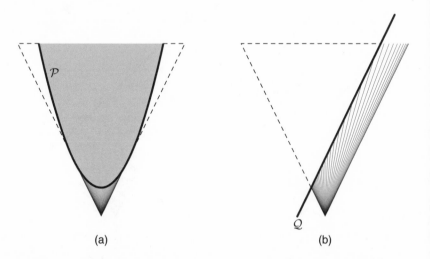

(a)   (b)

Figure 5.19. The parabola $\mathcal{P}$ is the intersection of a cone with a plane $\mathcal{Q}$ parallel to an element of the cone. The figure (a) is rotated by 90° in (b) showing the plane $\mathcal{Q}$ "on edge." The dashed lines in (a) and (b) indicate portions of the cone cut off by the plane.

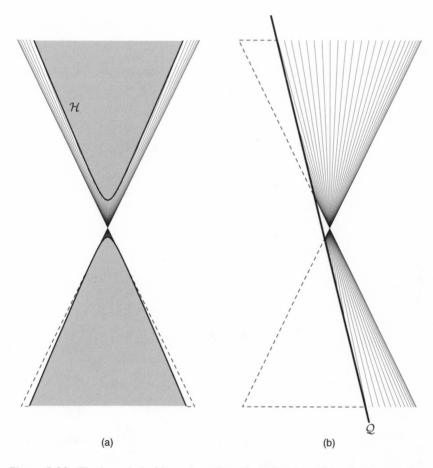

(a)   (b)

Figure 5.20. The hyperbola $\mathcal{H}$ consists of two branches formed by the intersection of the two nappes of the cone with a plane $\mathcal{Q}$ steeper than the elements of the cone. The figure (a) is rotated by 90° in (b) showing the plane $\mathcal{Q}$ "on edge." The dashed lines in (a) and (b) indicate portions of the cone cut off by the plane.

The conics are beautiful curves. I think that their appeal comes from the right mix of order and complexity. The suspension bridge is a beautiful structure partly because the main cable is in the shape of a graceful parabola. It is interesting that during the construction process, the cable is in the shape of another curve, the *catenary*—the shape of a suspended chain. It is only when the roadbed is attached that the horizontal uniform loading changes the shape of the cable to a parabola.

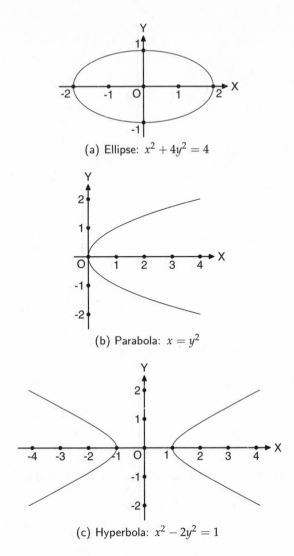

(a) Ellipse: $x^2 + 4y^2 = 4$

(b) Parabola: $x = y^2$

(c) Hyperbola: $x^2 - 2y^2 = 1$

Figure 5.21. Conic sections and their equations.

Coordinate representation of the conics

A very important technique of coordinate geometry is to associate curves with their equations. More specifically, a planar curve is associated with an equation involving two variables. We saw this in Figure 5.15, where the straight line, representing a motion of constant velocity, is associated with equation (5.4). More precisely, the relation between the line in Figure 5.15 and equation (5.4) is as follows:

A point with particular coordinates (t, s) in the S–T plane lies on the line in Figure 5.15 if and only if the numbers s and t satisfy equation (5.4). *A point lies on the curve if and only if the coordinates of the point satisfy the equation.*

Figure 5.22 is a second example showing the relationship between a curve and its equation. The equation of this circle is

$$x^2 + y^2 = 1 \qquad (5.5)$$

This means that the points with coordinates (x, y) that satisfy the equation $x^2 + y^2 = 1$ are precisely the points that belong to this circle. For example, the point (0.8,0.6) belongs to the circle because

$$0.8^2 + 0.6^2 = 0.64 + 0.36 = 1$$

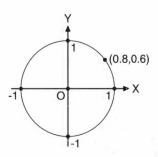

Figure 5.22. The equation of this circle is $x^2 + y^2 = 1$.

Equation (5.5) expresses the fact that the distance to the origin from any point on the circle is equal to 1. In fact, it follows from the Pythagorean theorem that the distance from the point (x, y) to the origin is equal to $\sqrt{x^2 + y^2}$.

Figure 5.2, showing the descent of a falling object, is an example of a parabola.

All of the conics are represented by equations. Here we ended our discussion by exhibiting three conics and their equations in Figure 5.21. However, the conics have much more structure—enough to fill out a course lasting several weeks. The easiest way to derive the structure of the conics is by using coordinate representations like Figure 5.21.

The visual sense is a wellspring of inspiration—in mathematics and in everyday life. The use of graphs in science and business, as well as mathematics, grew and flowered in the twentieth century to become universal.

Geometry has been guided by the visual sense since antiquity. But geometry gained certitude by the application of the axiomatic method. In the late Renaissance, coordinate geometry was created, supplementing the axiomatic method with algebraic reasoning. In the next chapters, we continue the story of algebra.

Part III

The Great Art

In this book, learned reader, you have the rules of algebra.

Written in five years, may it last as many thousands.

—Girolamo Cardano, Ars Magna (1545)

6

Algebra Rules

> Mephistopheles:
> *Dear friend, all theory is grey,*
> *And green the golden tree of life.*

> —JOHANN WOLFGANG VON GOETHE (1749–1832), Faust

THE ELOQUENT SEDUCTION of the above epigraph is difficult to resist. But remember that these are words of an agent of darkness — an adviser with questionable motives.

Mephistopheles certainly includes *algebra* when he speaks of *theory*. However, it is not he who moves many of our schoolchildren to say, "I hate algebra." We will not discuss the many and complex sources of the fear and hatred of algebra and mathematics in general; instead, we do hope to show that algebra is a useful and powerful *human* accomplishment. Algebra, far from being fearsome, is a comfortable place where we can confidently bring our most puzzling quantitative problems — a safe harbor for the mind.

It is certain that knowledge of elementary algebra is a first requirement for an increasingly long list of occupations — including many satisfying, challenging, and highly rewarding professions. In the fall of 2000, the lack of school preparation in algebra was considered so serious that the California legislature passed and the governor signed a law mandating the successful study of algebra as a requirement for graduation from high school.

What is algebra?

There is more than one kind of algebra. In Figure 6.1, we see the various branches of the algebra family tree. Here and in the rest of this chapter, *algebra* means *classical algebra*, the familiar school subject. Classical algebra is

111

distinguished from *abstract algebra,* an advanced mathematical topic concerned with certain axiomatically defined structures in which operations such as addition and multiplication are generalized or redefined. Abstract algebra is less than 200 years old.

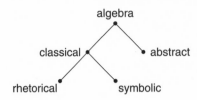

Figure 6.1. The algebra family tree.

Classical algebra is further divided into *rhetorical* and *symbolic* algebra. Symbolic algebra is the familiar school subject—a generalization of arithmetic using letters to stand for unknown or indefinite numbers in mathematical formulas and equations, together with certain methods of manipulation called the *rules of algebra.* Symbolic algebra in its present form is only about 400 years old—owing largely to the work of François Viète (1540–1603).

Classical algebra existed for thousands of years without the benefit of mathematical formulas. *Rhetorical algebra* is the statement and solution of complex problems of arithmetic *without the use of mathematical formulas.* In rhetorical algebra, problems are stated and solved in "prose." Examples of rhetorical algebra dating from about 1650 BCE are found in the ancient Egyptian Rhind Papyrus. The familiar "word problems" of school mathematics are stated rhetorically.

There are two gray areas in the above definition of rhetorical algebra:

1. It is arbitrary how complex an arithmetic problem must be to be called rhetorical algebra instead of just arithmetic.

2. With increased use of abbreviations and manipulative rules, rhetorical algebra gradually merges into symbolic algebra.

Algebra Anxiety

> *"When I use a word," Humpty Dumpty said in rather a scornful tone,*
> *"it means just what I choose it to mean—neither more nor less."*
> *"The question is," said Alice, "whether you can make words mean so many different things."*
> *"The question is," said Humpty Dumpty, "which is to be master—that's all."*
>
> —LEWIS CARROLL (1832–98), Through the Looking Glass

In the seventeenth century, there was a degree of suspicion and distrust of algebra—not because algebra was considered difficult but because some scholars felt that it lacked the rigor of geometry. For example, the philosopher Thomas Hobbes (1588–1679), who had more than a dabbling interest in mathematics, had an ongoing quarrel with the mathematician John

Wallis (1616–1703) over the merit of his recently published *Algebra* (1685). Hobbes referred to algebra as hen scratchings and a "scab of symbols."

Hobbes complained that the symbols of algebra either have no fixed meaning or are merely awkward abbreviations. Furthermore, Hobbes believed that, at best, algebraic symbols are a detour, needlessly forcing us to translate the mathematics back into ordinary English. Since Hobbes's accusations echo the feelings of many beginning students of algebra, it is worthwhile to discuss these criticisms.

It seems that Hobbes felt that algebra is a mathematical swindle—a shell game. The operator of a shell game shuffles three walnut shells by sliding them over a table. He challenges the victim to bet which shell contains the pea. The operator is generally quite skillful, and the victim generally loses his money. For Hobbes, it seems that the shells were the letters x, y, and z and the pea was mathematical truth. He feared the crafty algebraist would attempt to deceive by subtly switching meanings.

Hobbes was aware that algebra involves both abstraction and abbreviation, but he missed a crucial point. Once a problem is expressed algebraically, it can be transformed using the *rules of algebra*. These rules are based on ordinary logic. Once one has acquired some algebraic skill, the solution of a problem often reduces to a mechanical application of the rules of algebra. The solution is made easier by the fact that one need not translate the problem back to first principles or ordinary English until the end of a demonstration. Algebra has proved to be a reliable and successful labor-saving device for the mind.

Hobbes pioneered the idea that reasoning is a kind of verbal computation, but he does not seem to have considered the idea that reasoning might resemble algebraic computation. Today, it is easy to see the following pragmatic rebuttal to Hobbes's arguments: Algebra is the foundation for today's complex and successful technology.

Even Newton, the supreme mathematical genius of his time, had reservations about the legitimacy of algebra. How else can we explain the fact that his great work *Principia* (1687) is cast in geometric form when algebraic language would have been clearer and more natural? Did he omit mention of his greatest achievement, the calculus, because he considered it too algebraic? Several possible reasons have been suggested for Newton's reliance on geometry:

1. Newton felt that the matters that he discussed were truly geometric in nature.

2. Algebra was a controversial new science and geometry was an accepted ancient tradition. It seems that Newton was a follower of that tradition. Perhaps Newton felt that if he made his ideas geometric they would gain not only greater acceptance and understanding but also greater validity.

3. Perhaps Newton felt that algebra was a suitable tool for discovering

new ideas but not rigorous enough for presenting them—that an algebraic argument was merely *heuristic* and not logically rigorous.

Giants of mathematics—not schoolchildren only—have experienced algebra anxiety. There is an anecdote that relates to the discomfort many schoolchildren feel with elementary algebra:

> *Student:* Sir, I'm not comfortable with the passage at the bottom of page 6.

> *Professor:* Don't worry. When you have seen these words sufficiently often you will feel completely comfortable.

We may feel that the professor ought to give a specific explanation of matters that the student does not understand. Yet, algebra, for example, is sufficiently complex that the novice must learn to tolerate some gaps in understanding. When a professor studies a new research article, he also is the beginner, and he must initially tolerate some ambiguity and lack of understanding.

One unfamiliar with computers who embarks upon reading a computer manual has a similar experience. A page sometimes gives the impression that the reader is expected to know everything in the manual *except what is on that one page.* Good exposition helps, but only to a point with information that is strongly interconnected.

The beginning algebra student sometimes complains that an algebraic formula is not *real.* He or she needs to be assured that algebraic formulas *are* real. For the student, algebraic formulas attain reality gradually—like a foreign language—through accumulated successes.

Algebra enables us to solve problems using global patterns that do *not* have to be "understood" each time they are applied. Mathematics provides us with high-level principles that save us from the need to understand low-level details. George Pólya (1887–1985)—distinguished Hungarian-born American mathematician and celebrated writer on mathematical education—expressed this idea when he said, "Mathematics is being lazy. Mathematics is letting the principles do the work for you so that you do not have to do the work for yourself" (quoted in Walter and O'Brien (1986)).

Mathematicians use every possible labor-saving device to rachet up their understanding and to free themselves from details. For example, pencil and paper permits us to carry out computations that would overwhelm the unaided mind, and hand-held calculators do arithmetic with incredible speed and accuracy. Furthermore, larger digital computers execute calculations in seconds that would take eons with paper and pencil.

To use algebra effectively, we do not have to understand the first principles of what we are doing at every moment, but we need to know enough so that we are in control. We need to do better than the Wizard of Oz as he leaves helplessly in a hot air balloon: "I can't come back, folks! I don't know how it works! Goodbye!"[1]

There is a theory of teaching that suggests that the student should learn algebra the same way that it developed historically—following in days or weeks the route that humanity discovered through hundreds or thousands of years. It may reduce the anxiety of the beginning algebra student to know that for thousands of years there was rhetorical algebra *without algebraic formulas*.

Arithmetic by Other Means

Girolamo Cardano (1501–76) referred to the *rules of algebra* more than 400 years ago. However, the rules of algebra are principally the rules of *arithmetic*. Paraphrasing Baron von Clausewitz,[2] algebra is merely the continuation of arithmetic *by other means*.

In our early education in arithmetic, rules seem less important than the acquisition of computational skills—for example, by memorizing the multiplication tables. At this stage, the formula

$$11 \times (7 + 9) = 11 \times 7 + 11 \times 9 \tag{6.1}$$

is seen as a computational exercise—verified by computing the right side

$$11 \times (7 + 9) = 11 \times 16 = 176$$

and the left side

$$11 \times 7 + 11 \times 9 = 77 + 99 = 176$$

However, from the algebraic point of view equation (6.1) is an instance of the *distributive law of multiplication with respect to addition*:

For any numbers $a, b,$ and $c,$ we have $a(b + c) = ab + ac.$

The distributive law is one of about a dozen rules that could be listed explicitly. We have no need for an explicit statement here of all of these rules if we just keep in mind that the symbols $a, b, c,$ and so on, represent numbers and, therefore, can be manipulated as such.

In our first study of arithmetic, numbers are like our circle of family and friends: We know everyone's name, and we know the relationship of each person to everyone else. Later, when we study algebra, we are the general of a great anonymous army of numbers. To the great benefit of the strategy and tactics of algebra, numbers follow rules with greater obedience than do real soldiers.

The general issues orders: "Every soldier will be issued a helmet liner." In algebra, we also make general statements: "For every number $x, x + x =$

$2x$." "There exists a number y such that $3y + 10 = 19$." In mathematical logic, the phrases "for every" and "there exists" are called *quantifiers*. They occur with such frequency in mathematics that there are standard abbreviations: ∀ and ∃, respectively — however, we will not make further use of these symbols.

Symbolic algebra

My Dear Aunt Sally

Figure 6.2.

In the above paragraphs, we are using symbolic algebra when we use the letter x to represent an unknown number. Symbolic algebra also uses special symbols and conventions to represent arithmetic operations. For example, ab is the product of the numbers a and b. There is an important convention for interpreting algebraic (and arithmetic) formulas known as the *order of precedence* that requires that multiplications must be done first, then divisions, additions, and subtractions, in that order. The mnemonic for this convention is "My Dear Aunt Sally" (Figure 6.2). This rule is arbitrary but used universally. Computer languages and electronic calculators generally implement this order of precedence.[3]

For example, in the expression $5 \times 3 + 4 \div 2 - 1$, first multiply ($5 \times 3 = 15$), then divide ($4 \div 2 = 2$), then add ($15 + 2 = 17$), and finally subtract ($17 - 1 = 16$). The order of precedence can be overridden using parentheses. The expressions in inmost parentheses are always computed first. For example:

$$5 \times (3 + 4) \div (2 - 1) = 5 \times 7 \div 1 = 35$$

Symbolic algebra is largely concerned with *equations*. An equation, like $x + x = 2x$, that is true for all numbers x is called an identity. To emphasize that an equation is an identity, sometimes the symbol $\equiv$ is used instead of the usual equal sign (e.g., $x + x \equiv 2x$). On the other hand, $3x + 10 = 19$ is *not* an identity because it is true only for one particular value of x.

Strategy and tactics. A first course in algebra is devoted largely to the problem of *solving* an equation — finding the values of the unknown for which the equation is true. A solution is accomplished if we can find a chain of equations with the following properties:

- The first equation of the chain is the given equation.
- Each further equation is equivalent to the one that precedes it.
- The last equation of the chain consists of x on the left side of the equal sign and a particular number, the solution, on the right side (e.g., $x = 3$). A solution of an equation is also called a *root* of the equation.

Two simple methods of transforming an equation into an equivalent one are as follows:

1. Add (or subtract) the same quantity to both sides of the equation.
2. Multiply (or divide) both sides of the equation by the same nonzero quantity.

Starting with the equation $3x + 10 = 19$:

1. Subtract 10 from both sides, obtaining $3x = 9$.
2. Divide both sides by 3, obtaining the solution $x = 3$.

The strategy consisted of two steps:

1. Add (or subtract) a quantity to both sides so that the left side is a multiple of x, and the right side does not involve x.
2. Multiply (or divide) both sides by a quantity so that the transformed equation asserts that x is equal to a particular number — not depending on x.

This simple strategy works for Examples 6.1–6.3, but some problems are more difficult. The examples below are a few word problems and their solutions in rhetorical and symbolic algebra. For the simplest problems, the rhetorical solution seems sufficient and the symbolic solution unnecessary. For more difficult problems, the rhetorical solution tends to require a different clever idea for each problem, whereas the symbolic solution uses standard methods.

Our first problem — more than 3600 years old — is from the ancient Egyptian Rhind Papyrus (see Chapter 1).

Example 6.1 (Rhind Papyrus, Problem 24). A quantity plus one-seventh of it becomes 19. What is the quantity?

Solution (rhetorical, modern version). We eliminate the reference to the fraction one-seventh by restating the premise of the problem as follows: *Seven times a quantity plus the quantity itself is equal to* 19×7. In other words, 8 times the quantity is equal to $19 \times 7 = 133$. Thus, we see that this quantity must be 133 divided by 8, that is $16\,5/8$.

The above problem is an easy exercise for beginning algebra students; however, it was not such an easy problem for the Egyptians because they used the awkward system of unit fractions described in Chapter 1.

Solution (rhetorical, Egyptian style). First guess: Maybe the unknown quantity is 7. (This number is chosen because the calculation is easy.) But this fails because $1\,1/7$ of 7 is 8, not 19.

Next step: The correct answer must be $19/8$ times 7. (In Egyptian notation, $19/8$ must be written $2 + 1/4 + 1/8$.) Multiplying by 7, the Egyptians obtain:

$$7 \times 2 = 14$$
$$7 \times 1/4 = 1 + 1/2 + 1/4$$
$$7 \times 1/8 = 1/2 + 1/4 + 1/8$$

Adding the right sides of these equalities yields:
$$7 \times (2 + 1/4 + 1/8) = 14 + (1 + 1/2 + 1/4) + (1/2 + 1/4 + 1/8)$$
$$= 16 + 1/2 + 1/8$$

Finally, we give a symbolic demonstration:

Solution (symbolic). Let x be the required quantity. The problem states

$$x + x/7 = 19$$

The left side of this equation is equal to $\frac{8}{7}x$. Multiplying both sides of the equation by $7/8$, we obtain

$$x = 19 \times \frac{7}{8} = \frac{7 \times 19}{8} = 16\,5/8$$

Example 6.2. In 3 years Ada will be twice as old as she was 7 years ago. How old is she now?

Solution (rhetorical). Her age doubles from the time 7 years ago to 3 years from now — a period of $7 + 3 = 10$ years. This implies that she was 10 years old 3 years ago and she will be 20 in 3 years. Therefore her current age is 17.

Solution (symbolic). Since many symbolic solutions follow this pattern, we supply all the relevant details.

When we approach such a problem, in general, we do not know how many solutions there are, or if there are any solutions at all. We begin by expressing the conditions of the problem by means of an algebraic equation.

Let x be Ada's current age. The conditions of the problem are met if there exists a number x such that

$$x + 3 = 2(x - 7) \tag{6.2}$$

We say that such a number is a *solution*, or that it *satisfies*, the equation. Thus, we have transformed the original problem concerning Ada's age to the algebraic problem of finding a solution of an equation.

Our plan is to use rules of algebra to find a sequence of equivalent equations terminating with an equation of the form $x =$ some number.

The *rules of algebra* are of two types:

1. Methods of simplifying or expanding an algebraic expression. For example, the *distributive law of multiplication* can be applied to the right side of equation (6.2) to obtain the $2(x - 7) \equiv 2x - 14$.
   A second method (among others of this type) is called *collecting like terms*. For example, by collecting like terms, the expression

   $$x + 3 + 14 - x$$

   is identical to 17.
2. Methods of transforming an equation into an *equivalent* equation with exactly the same solutions. For example, the rules of algebra permit us to add or subtract the same quantity from both sides of the equation, or to multiply both sides by the same nonzero quantity. Adding $14 - x$ to both sides of equation (6.2) gives

   $$x + 3 + 14 - x = 2x - 14 + 14 - x \qquad (6.3)$$

   Collecting like terms on both sides of this equation, we obtain the equivalent equation $17 = x$.

The final step is to check that 17 satisfies the conditions set down for Ada's age. (It is necessary to check, e.g., the unspoken requirement that Ada's age is a nonnegative integer that is not implausibly large.)

Example 6.3. The hare was so bored racing the tortoise that he decided to take a nap 1 mile from the finish line. When he awoke, the tortoise, still plodding at 0.2 mile per hour, had only 52.8 feet (0.01 mile) to go. The hare started running full speed at 35 miles per hour and just barely won the race. How far from the finish were they and what time was it when the hare passed the tortoise?

Solution (rhetorical). The distance between the tortoise and the hare is initially $1.00 - 0.01 = 0.99$ mile and decreases at $35.0 - 0.2 = 34.8$ miles per hour. Therefore the tortoise and the hare meet when the elapsed time is $0.99 \div 34.8$, approximately 0.02845 hour (1.707 minutes). In that time the hare has traveled $0.02845 \times 35.0 = 0.9957$ mile. Since the initial distance of the hare from the finish line was 1.0 mile, he has not yet reached that point. In fact, the hare overtakes the tortoise at approximately $1.0000 - 0.9957 = 0.0043$ mile, that is, 22.7 feet before crossing the finish line.

Solution (symbolic). Let t be the time in hours since the hare awoke and started running. The distances, in miles, of the hare and the tortoise from the finish line are $1 - 35t$ and $0.01 - 0.2t$. At the moment that the hare overtakes the tortoise these distances are equal; that is, $1 - 35t = 0.01 - 0.2t$. Adding $35t - 0.01$ to both sides of this equation yields $0.99 = 34.8t$. Now dividing both sides by 34.8, we obtain $t = 0.99 \div 34.8 \approx 0.02845$, where t is the moment at which the hare overtakes the hare. The distance in miles from the finish line at this moment is approximately $1 - 35 \times 0.02845 = 0.00425$. Since this number is positive, the hare overtakes the tortoise *before* crossing the finish line. Hence, the hare wins the race.

Equations are an important tool for solving problems like the three above. With a new tool there are two natural courses of action:

1. Use the tool to dig deeper.
2. Sharpen and improve the tool.

The following sections survey some of the achievements from pursuing these two strategies.

Algebra and Geometry

Figure 6.3. Division of the rectangle $\mathcal{R}$.

The algebra of the ancient Egyptians and Babylonians consisted of recipes for certain calculations—without proofs. The Greeks were the first to give algebraic proofs—in Book II of Euclid's *Elements*. However, Euclid's proofs are entirely geometric, and we need to reinterpret these theorems to see that they a concerned with algebra. For example, *Proposition 1*, illustrated in Figure 6.3, asserts that *if a rectangle $\mathcal{R}$ is divided by a line segment parallel to one of its sides into two rectangles $\mathcal{S}$ and $\mathcal{T}$, then the area of $\mathcal{R}$ is equal to the sum of the areas of $\mathcal{S}$ and $\mathcal{T}$.*[4]

Euclid's proposition may seem entirely geometric, but, in fact, it is an instance of rhetorical algebra. The true nature of the proposition becomes evident if we translate it into an algebraic formula. If a, b, and c are the lengths of the sides of the rectangles, as shown in Figure 6.3, the proposition asserts $a(b + c) = ab + ac$, which is the *distributive law of multiplication with respect to addition*. Euclid's proposition is an instance of rhetorical algebra.

As we will see in the next section, geometric arguments can enhance an algebraic argument. However, before the advent of symbolic algebra at the end of the sixteenth century, such geometric arguments were considered indispensable. This dependence of algebra on geometry might be explained by the great prestige and authority enjoyed by Euclidean geometry, but this fails to explain why geometric algebra faded when symbolic algebra appeared. The more compelling reason is that symbolic algebra provided more natural and convincing algebraic arguments.

An algebraic argument can be supported either by a geometric figure or an algebraic formula. Where one used to say, "Behold, the figure," now one says, "Behold, the formula."

Al-jabr

The word *algebra* comes from *Hisab **al-jabr** w'al-muqabala*, the title of a work by the Baghdad mathematician Abu Jafar Muhammad ibn Musa al-Khwarizmi (780?–850?). (The word *algorithm* comes from the name al-Khwarizmi.) In that work, al-Khwarizmi showed, for example, a solution of the quadratic[5] equation

$$x^2 + 10x = 39 \qquad (6.4)$$

al-Khwarizmi's algebra is rhetorical. The terms x^2 and $10x$ he calls "a square" and "10 roots," respectively. He writes:

> A square and 10 roots are equal to 39 units. The question there-fore in this type of equation is about as follows: what is the square which combined with ten of its roots will give a sum total of 39? The manner of solving this type of equation is to take one-half of the roots just mentioned. Now the roots in the problem before us are 10. Therefore take 5, which multiplied by itself gives 25, an amount which you add to 39 giving 64. Having taken then the square root of this which is 8, subtract from it half the roots, 5 leaving 3. The number three therefore represents one root of this square, which itself, of course is 9. Nine therefore gives the square. (Grant (1974, p. 111))

In modern terms, this solution involves adding 25 to each side of the equation to "complete the square."

$$x^2 + 10x + 25 = 64 \qquad (x+5)^2 = 8^2 \qquad x + 5 = 8 \qquad x = 3 \qquad (6.5)$$

(Since al-Khwarizmi did not know negative numbers, he missed the root $x = -13$.)

Al-Khwarizmi felt that it was necessary to sup-plement the foregoing solution with a geometric version. His explanation is ambiguous. The follow-ing is a modern interpretation (see Parshall (1995)):

In Figure 6.4, the darkly shaded square has side x and area x^2, and each of the $x \times 5$ lightly shaded rectangles have area $5x$. Thus, the total area of the shaded L-shaped region is $x^2 + 10x$, the left side of equation (6.4). Appending the unshaded square of area $5 \times 5 = 25$, we obtain the entire large square in Figure 6.4. Since equation (6.4) requires that the L-shaped region have area 39, it follows that the entire square has area $39 + 25 = 64$. Since $\sqrt{64} = 8$, the side of

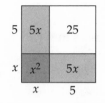

Figure 6.4. Solving a quadratic equation.

the large square must be 8. On the other hand, the side is equal to $x + 5$. Hence $x + 5 = 8$, and x must be equal to 3.

The following word problem leads to quadratic equation:

Example 6.4 (The valentine exchange). Ada teaches a third grade class in which every child gives a valentine card to every other child. Ada used exactly 380 valentine cards for this project. How many children were in the class?

Suppose that there are n children in the class. Each child gives a card to each of the $n - 1$ other children. Since there are n children, the total number of cards sent is $n(n - 1)$. To find n, we must solve the equation $n(n - 1) = 380$.

Solution (Trial and error). We compute $n(n - 1)$ for various choices of n. We find $20 \times 19 = 380$, which shows that there must be 20 children in the class.

Al-Khwarizmi's method gives a second method of solving this equation. However, there is no benefit in using his geometric explanation. This method, completing the square, can be used to solve any quadratic equation.

Solution (Completing the square). To solve the equation $n(n - 1) = 380$, we add $1/4$ to both sides. This makes the right side a perfect square:

$$n(n - 1) + 1/4 \equiv n^2 - 2n + 1/4 \equiv (n - 1/2)^2 = 3801/4$$

Taking square roots of both sides, we obtain

$$n - 1/2 = \sqrt{3801/4}$$

and adding $1/2$ to both sides

$$n = \sqrt{3801/4} + 1/2$$

We have reduced the problem to finding the square root of $3801/4$. In fact, for any quadratic equation, the method of completing the square reduces the problem to the computation of a square root. The next section discusses methods for calculating square roots, but here we use a calculator to find $\sqrt{380.25} = 19.5$, so that our solution is $n = 19.5 + 0.5 = 20$.[6]

The next section discusses two methods of computing square roots. One method used to be a standard topic of instruction in U.S. schools, and the other stems from the ancient Babylonians.

Square root algorithms

The following type of problem is a concern that undoubtedly dates from the beginnings of agriculture — the division of land.

Problem 6.1. Find the dimensions of a square field with area equal to 14,000 square feet.

Our modern solution notes that the length of the side in feet is the square root of 14,000. If we have a calculator or a set of mathematical tables, we quickly find $\sqrt{14,000} = 118.32\ldots$. Otherwise, we could use the numerical algorithm resembling long division, shown in Figure 6.5(a), that used to be a standard part of eighth grade mathematics.

The "standard" square root algorithm

The square root algorithm has been dropped from the curriculum of most schools. Not every student enjoyed the careful attention to detail required by this computation. Nevertheless, my experience in school would have been diminished if this topic had been dropped before my time. I found it fascinating that this magical sequence of calculations could produce the square root of a number and that I could verify its correctness with ordinary multiplication. This was my first exposure to an algorithm of such power.

Some educators argue that algorithms for square roots, and for even long division, are no longer needed because such calculations are easily done today on hand-held calculators. Nevertheless, the concept of algorithm is the foundation of modern computer technology. I believe that school education should include the experience of carrying out detailed computations with pencil and paper or—even better—with a *computer spreadsheet.*[7]

The geometric motivation for the square root algorithm is illustrated in Figure 6.5(b). Very roughly, the idea is as follows: Start with a square whose side is an approximation of the desired square root, for example, the square with the dark shading in Figure 6.5(b). Find a better approximating square by appending an L-shaped region—shown with lighter shading in Figure 6.5(b). Repeat until a sufficient approximation of the desired square root is obtained. More precisely, the following is a line-by-line explanation of the algorithm carried out in Figure 6.5(a):

A. We wish to compute the square root of 14,000. The first step is to group the digits of this number by twos, as shown.

B. Find the largest one-digit number whose square does not exceed the leftmost group of A. Write this number on the top line as the first digit of the square root, and write its square $(1^2 = 1)$ as the leftmost entry on line B. Enter a 1, the first digit of the square root, on the top line. In this step we obtain 100 as the first approximation of $\sqrt{14,000}$. This corresponds to the darkly shaded square in Figure 6.5(b). This square represents an area of 10,000 square feet; we are looking for a square of area 14,000 square feet.

C. Subtract row B from the first two groups of row A, obtaining 40. In Figure 6.5(b), this number represents the fact that the area of the dark shaded square is 4,000 square feet too small.

D. In this row, we find that a square of side 110 gives the second approximation. In Figure 6.5(b), this approximation is obtained by augmenting the dark shaded square with the lighter shaded L-shaped region consisting of a small square and two rectangles. These geometric steps correspond in Figure 6.5(a) to the following algorithmic steps:

(a) Double the portion of the square root already found, that is, double 1, obtaining 2. (Doubling is necessary to account for *two* rectangles of the L-shaped region.)

(b) Form a two-digit number by appending a digit d on the right side of 2, the number found above in (a), obtaining the number $20 + d$. The digit d is found by trial so that $d \times (20 + d)$ is as large as possible, but not exceeding 40, the remainder found in C.

(c) Enter 1 (the digit d) on the top line as the second digit of the square root.

(d) The entry 21 on line D represents the fact that the area of the lightly shaded L-shaped region in Figure 6.5(b) is 2100 square feet.

E. Subtract $(40 - 21 = 19)$ and bring down a pair of zeros.

F. As in D, double the portion of the root already found: $2 \times 11 = 22$. Form a three-digit number by appending a digit d on the right side of 22. The digit d is found by trial to be 8 by the requirement that $d \times (220 + d)$ is as large as possible but not exceeding 1900. Enter the product: $8 \times 228 = 1824$.

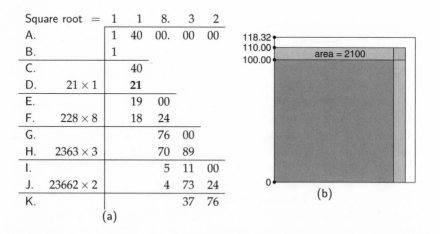

Figure 6.5. An algorithm for computing square roots and its geometric interpretation.

G. As in E, subtract $(1900 - 1824 = 76)$ and bring down a pair of zeros. The remaining steps (H through K) follow the pattern established in D through G.

The Babylonian square root algorithm

The ancient Babylonians had a method of approximating the square root of a number n. Start with a rough guess r for the square root. If this guess is correct, then $r = \frac{n}{r}$. If r is too small or too large then $\frac{n}{r}$ is, respectively, too large or too small. Hence, the average of r and $\frac{n}{r}$ is a better approximation for the square root than either of these two numbers. This leads to the approximation:

$$\sqrt{r} \approx \frac{r + \frac{n}{r}}{2}$$

For example, the approximate the square root of $n = 14,000$, we might use the guess $r = 100$, since $100^2 = 10,000$. We would obtain

$$\sqrt{14,000} \approx \frac{100 + 14,000/100}{2} = 50 + 70 = 120$$

The square root of 14,000 is 118.3215957..., which gives the Babylonian approximation an error of 1.4%, good enough for most everyday early Sumerian uses.

It might have occurred to some clever Babylonian that he or she could get an even better approximation by starting with the above approximation as the initial guess, that is, by putting $r = 120$, obtaining an approximation with an error of only 0.01% — a stunning improvement:

$$\sqrt{14,000} \approx \frac{120 + 14,000/120}{2} = 118\frac{1}{3}$$

The next step is an infinite sequence of *successive approximations* for the square root. (The Babylonians did not carry matters this far.) We use the result of each approximation as the guess for the next iteration. That is, starting with r_0 as an initial guess for the square root of n, we build the following sequence of approximations:

$$r_1 = \frac{r_0 + \frac{n}{r_0}}{2}$$

$$r_2 = \frac{r_1 + \frac{n}{r_1}}{2}$$

$$r_3 = \frac{r_2 + \frac{n}{r_2}}{2}$$

$$\cdot \quad \cdot \quad \cdot$$

For $n = 14,000$ and $r_0 = 100$, we have computed r_1 and r_2 above. Continuing one more step, we find:

$$r_4 = \frac{118\text{\textonehalf}/3 + \frac{14,000}{118\text{\textonehalf}/3}}{2} = 118.32159624\ldots$$

Since the true value of $\sqrt{14,000}$ is $118.32159566\ldots$, r_4 has an error of only 0.0000005%, far beyond what is needed for building most ziggurats.

Figure 6.6 illustrates the use of the Babylonian square root algorithm to compute $\sqrt{14,000}$. We start with two graphs, the curve

$$C : s = \frac{r + \frac{14,000}{r}}{2}$$

and the diagonal straight line D ($s = r$). For any point (r, s) on C, s represents the succeeding Babylonian approximation of $\sqrt{14,000}$ starting from r as an initial guess.

The point Z is the intersection of the curves C and D. Let R be the (horizontal) r-coordinate of Z. We see that if we started with R as an initial guess for $\sqrt{14,000}$, the succeeding Babylonian approximation would also be R; in other words, R is precisely the square root of 14,000.

Of course, we cannot expect to be so lucky as to choose R as our initial guess. (In Figure 6.6, the initial guess r_0 happens to be equal to 40.) Our iteration follows the arrows in the figure:

1. We proceed upward from the first guess r_0 on the (horizontal) r-axis to the point A on C.

2. The (vertical) s-coordinate of A, r_1, is the second approximation. To obtain the third approximation, we must locate a point with (horizontal)

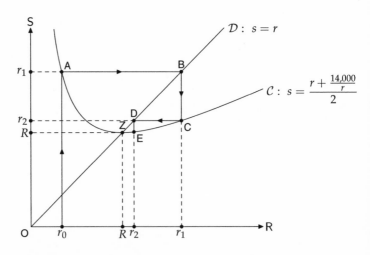

Figure 6.6. The Babylonian square root algorithm.

r-coordinate equal to r_1. This is done by proceeding horizontally to point B on $\mathcal{D}$.

3. To find the third approximation, r_2, proceed vertically from B to C on $\mathcal{C}$. The (vertical) s-coordinate of C is equal to r_2.

4. As in step 2, we proceed horizontally to D on $\mathcal{D}$. The (horizonal) r-coordinate of D is r_2.

5. As in step 3, proceed vertically to E whose (vertical) r-coordinate gives us the fourth approximation. And so on.

We see that the path ABCDE winds ever closer to the point Z. The Babylonian algorithm converges very rapidly to the square root. In fact, it can be shown that at each iteration the number of correct digits of the approximation tends to double. The Babylonian algorithm is clearly a more powerful square root algorithm than the standard one.

The Babylonian square root algorithm is illustrative of the modern iterative methods of numerical analysis. Kreith and Chakerian (1999) show how this and other iterative algorithms are particularly suitable for inclusion in today's school curriculum.

We have seen some of the early beginnings of algebra. Next, we look in more detail at the problem of solving equations—leading to rambunctious quarrels between certain Italian mathematicians of the sixteenth century.

7

The Root of the Problem

Algebra is generous; she often gives more than is asked of her.

—JEAN LE ROND D'ALEMBERT (1717–83)

AY "ALGEBRA" and one thinks "equations." Indeed, equations are a central concern of classical algebra. An equation poses a mathematical "whodunit." The culprit x—the *unknown* x—left clues in the form of an equation, and the mathematical detective must determine its identity, the *root* (or roots) of the equation. As we will see, for certain Italian mathematicians of the sixteenth century, equations led to high intrigue, but probably no actual crimes.

— What in the world is $x = y$? How can I make any sense of that gibberish?
— This is it.
— Yes, go on. This is what?
— Without a context, neither the English pronouns, 'this' and 'it,' nor the mathematical pronouns, x and y, make sense.

The terms *square root*, *cube root*, and so on, refer to roots of certain particular equations. For example, the square and cube roots of 2 ($\sqrt{2}$ and $\sqrt[3]{2}$) denote the positive roots of the equations $x^2 = 2$ and $x^3 = 2$, respectively.

To find a root of the equation

$$x^3 + 6x + 2 = 0 \qquad (7.1)$$

means to find a numerical value for x that makes the equation an identity. That seems clear enough, but it is not. Indeed, there are three different ways to understand what it means *to find a value for* x:

1. Find a rational number x that solves the equation exactly.

128

2. Use rational numbers to approximate x to any desired degree.

3. Express x in terms of rational operations (addition, subtraction, multiplication, and division) and *radicals* (square roots, cube roots, and so on).

1. *Find a rational solution.* This is not an appropriate requirement because many equations involving a single unknown x do not have rational solutions even though rational numbers can approximate a solution as closely as desired. However, there is a type of problem involving two or more unknowns, in which one looks only for rational solutions. For example, the equation $5x - 3y = 1$ has a solution $x = 2$, $y = 3$. Such problems are called *Diophantine equations* after Diophantus of Alexandria (fl. 250? CE), who solved a great many problems of this type. We will not pursue this sort of problem in this chapter.

2. *Use rational numbers to approximate a solution.* A solution in this sense satisfies the needs of current technology. Graphical methods, discussed below, and iterative methods like the Babylonian square root algorithm (page 125) lead to this kind of solution.

3. *Express the solution in terms of rational operations and radicals.* For example, we will see later $x = -\sqrt[3]{4} + \sqrt[3]{2}$ is a root of equation (7.1). From rational approximations we might never learn that a solution could be expressed using cube roots. On the other hand, if we need the solution for some practical purpose, we still must to find decimal approximations for the cube roots.

From our modern perspective, item 2 seems the natural way to find the solution of an equation. However, for mathematicians of the Italian Renaissance, item 3 seemed the only sensible way to understand the solution of an equation—mere approximations probably would have been unacceptable. In this chapter, we consider the solution of equations mainly in the sense of item 3. Nevertheless, the next section deals with graphical solutions.

This chapter continues the discussion of equations that began in the preceding chapter, for example, linear equations in Examples 6.1–6.3 and a quadratic equation in Example 6.4. In this chapter, we will also look at cubic equations.

We begin with some graphical solutions.

Graphical Solutions

The following three examples illustrate the graphical solution of equations:

$$x + \frac{x}{7} = 19 \qquad \text{see Example 6.1.} \qquad (7.2)$$

$$x^2 + 10x = 40 \tag{7.3}$$

$$x^3 - 100x + 200 = 0 \tag{7.4}$$

To solve one of these equations means to find a numerical value for x that makes the equation an identity. We prepare for the graphical solution by introducing the equations that also involve a variable y (see Figures 7.1 (a)–(c), respectively.):

$$y = x + \frac{x}{7} - 19 \tag{7.5}$$

$$y = x^2 + 10x - 40 \tag{7.6}$$

$$y = x^3 - 100x + 200 \tag{7.7}$$

A solution of equation (7.2), (7.3), or (7.4), is a value of x such that the corresponding value for y is 0 in equations (7.5), (7.6), or (7.7). Graphs of these equations are shown in Figure 7.1. The problem of solving the three equations (7.2), (7.3), and (7.4), that is, finding their roots, is the same as the problem of finding the points A, B, C, D, E, and F where the graphs in Figure 7.1 cross the horizontal axes.

Roots of equations found by the graphical method are only approximate. Greater accuracy depends on the precision of our draftsmanship. One way to improve accuracy is to draw the graph only in the neighborhood of the crossing of the horizontal axis. Figure 7.1(a-i) is a normal scale graph for equation (7.5); the region enclosed in the small rectangle about the crossing point A in this figure is shown magnified in Figure 7.1(a-ii). With this magnification we can see that A, the root of equation (7.5), is between $x = 16.62$ and $x = 16.63$. From the solution of Example 6.1, we know the exact root of this equation is $x = 16.625$.

Similarly, Figures 7.1(b-ii) and (c-ii) show magnifications of the larger scale graphs shown in Figures 7.1(b-i) and (c-i). We see that C, one of the two roots of equation (7.6), is approximately $x = 3.05$; and F, one of the three roots of equation (7.4), is approximately $x = 8.8$. Using nongraphical methods that will be discussed later, the roots of these two equations can be found to any desired accuracy. In fact, $x = 3.06226$ and $x = 8.78885$ are approximations accurate to five decimal places of the roots C and F, respectively.

Figure 7.1(a-i), the graph of equation (7.5), is a straight line. The magnified graphs Figures 7.1(b-ii) and (c-ii) appear almost straight. The greater the magnification, the straighter they will appear.

Quadratic Equations

In this section, we consider quadratic equations in more detail. We begin with an example: the *golden section*, a proportion that can be found in the

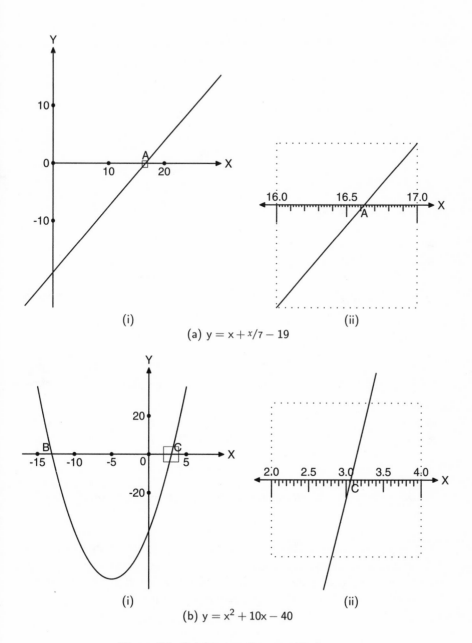

(i)   (ii)

(a) $y = x + x/7 - 19$

(i)   (ii)

(b) $y = x^2 + 10x - 40$

Figure 7.1. Solving equations graphically.

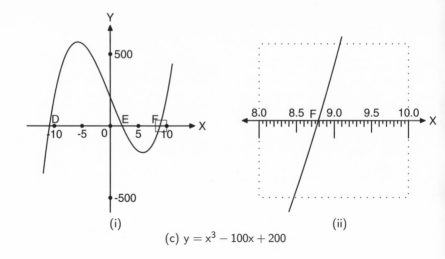

(c) $y = x^3 - 100x + 200$

Figure 7.1. (*continued*) Solving equations graphically.

construction of the Parthenon. The golden section was also used by the twentieth-century architect LeCorbusier.

Example 7.1 (the golden section). The ancient Greeks believed that a rectangle is most beautiful if the ratio of the larger side to the smaller equals the ratio of the sum of the sides to the larger. If the length of the longer side is x times the length of the smaller, find the value of x that achieves the golden section.

Solution (completing the square). Referring to Figure 7.2, we must have the proportion $x : 1 :: x + 1 : x$, or, in other words,

$$x = \frac{x+1}{x}$$

Multiplying both sides of this equation by x, we obtain the quadratic equation

$$x^2 = x + 1 \tag{7.8}$$

The key step of this solution is an algebraic technique called *completing the square*, which transforms equation (7.8) so that the left side is a perfect square and the right side is a constant independent of x. To achieve this, we add $1/4 - x$ to both sides of this equation, obtaining

$$x^2 - x + 1/4 = (x + 1/2)^2 = 5/4$$

Taking square roots of both sides of this equation, we obtain $x - 1/2 = \pm 1/2\sqrt{5}$, which is equivalent to

$$x = \tfrac{1}{2}(1 \pm \sqrt{5})$$

Since x represents the longer side of the rectangle, we must choose the plus sign:

$$x = \tfrac{1}{2}(1 + \sqrt{5}) \tag{7.9}$$

Using one of the square root algorithms (or an electronic calculator), we find $\sqrt{5} \approx 2.236068$, yielding

$$x \approx 1.618034 \tag{7.10}$$

Which is the true answer to this problem, equation (7.9) or equation (7.10)? The answer depends on our point of view. For the builders of the Parthenon, for me when I drew Figure 7.2, and for anyone interested in measuring or building, equation (7.10) is the useful answer. On the other hand, equation (7.9) displays what mathematicians would call the number theoretic properties of the golden section. It shows that the golden

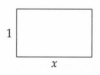

Figure 7.2. The golden section.

section has a rational[1] part $u = 1/2$ and an irrational part $\sqrt{5}/2$. It can be shown that the square root of any rational number is irrational provided that it is not a perfect square.

This idea can be used to construct a second solution of Example 7.1, that is, a second method for solving equation (7.8). This method is important because it is very similar to the method for solving cubic equations that will be discussed further below. The idea of the method is to guess the form of the solution and then to show that, indeed, there exists a solution of that form. What would be an acceptable guess?

Of course, it would not be an acceptable solution to guess that formula (7.9) is a root of equation (7.8) and then to give a calculation that shows that this guess is correct. This procedure is unsatisfactory because this "guess" is not plausible.

However, it is plausible to guess that a solution of equation (7.8) can be achieved by a formula that involves nothing worse than a square root. Perhaps one might guess that the solution is of the form $a + b\sqrt{5}$ where a and b are rational numbers. This is better, but it still seems unlikely to guess that $\sqrt{5}$ is involved.

It is more plausible to guess that the solution is of the form $a + b\sqrt{D}$ where a and b and D are rational numbers and D is not a perfect square. If D is fixed and $a + b\sqrt{D} = 0$, then a and b must both be equal to zero. Generalizing this idea, let us look for the solution of equation (7.8) in a universe of numbers of the form $u + v$, numbers that can be expressed as a sum of a "good" part u and a "bad" part v. We assume that, like the numbers of the form $a + b\sqrt{D}$, if the number $u + v$ is equal to zero then both u and v must also be equal to zero. It follows that zero is the only number that is simultaneously good and bad.

The multiplication of good and bad numbers satisfies Table 7.1. The motivation, for example, for assuming that the product of two bad numbers is good is that the product of two rational multiples of $\sqrt{D}$ is a rational number. This idea leads to the following alternate solution for Example 7.1.

Table 7.1. Good/bad table of multiplication.

	good	bad
good	good	bad
bad	bad	good

Solution (good and bad parts). Assume that there is a solution of equation (7.8) of the form $u + v$ where u is good and v is bad. In equation (7.8), we replace x by $u + v$:

$$(u + v)^2 \equiv u^2 + 2uv + v^2 = (u + v) + 1$$

We rearrange this equation by putting all the good terms on the left side and the bad ones on the right. (The only bad terms are the ones with an odd power of v as a factor.)

$$\underbrace{u^2 + v^2 - u - 1}_{\text{good}} = \underbrace{v - 2uv}_{\text{bad}} \tag{7.11}$$

Since only 0 is both good and bad, we expect to solve equation (7.11) so that both sides are equal to zero. If we can do so, then we are finished because now $x = u + v$ becomes a solution of equation (7.8). Note that the validity of this conclusion does not depend on Table 7.1 or any other of our assumptions concerning good and bad numbers. Setting the right (bad) side equal to zero, we have $v(1 - 2u) = 0$. This implies that either v is zero or $u = 1/2$. Let us try the latter alternative. If $u = 1/2$, then, setting the left (good) side of equation (7.11) equal to zero, we have

$$u^2 + v^2 - u - 1 = 1/4 + v^2 - 1/2 - 1 = 0 \tag{7.12}$$

Solving for v, we have $v = \pm 1/2\sqrt{5}$. As before, we find the solutions

$$x = \tfrac{1}{2}(1 \pm \sqrt{5})$$

(The second alternative, $v = 0$, does not produce any further solutions.)

The quadratic formula. Either of the above two methods for solving Example 7.1 can be used to solve the general quadratic equation:

$$ax^2 + bx + c = 0 \tag{7.13}$$

(Note that a, b, and c are constants.) The quadratic formula gives two solutions:

$$x = \frac{-b \pm \sqrt{b^2 - 4ac}}{2a} \tag{7.14}$$

1. If $b^2 - 4ac$ is positive, then equation (7.13) has two distinct real solutions.

2. If $b^2 - 4ac$ is zero, then these two solutions coalesce into one: $x = -b/2a$.

3. If $b^2 - 4ac$ is negative, then equation (7.13) has solutions involving the square root of a negative number. This was a very puzzling idea in the sixteenth century when such roots were dubbed *imaginary*. In particular, the number $\sqrt{-1}$ is called the imaginary unit and is denoted i. A number of the form $a + ib$, where a and b are ordinary real numbers, is called a *complex number*. The complex numbers are completely legitimate and useful in both pure and applied mathematics.

Secrecy, Jealousy, Rivalry, Pugnacity, and Guile

Not the fifth circle of hell or an evil legal firm — rather, five Italian mathematicians of the sixteenth century are the protagonists of our story. Their names are Scipione del Ferro (1465–1526), Antonio Maria Fior (born 1506), Niccolò Fontana (Tartaglia) (1500?–57), Ludovico Ferrari (1522–65), and Girolamo Cardano (1501–76). Their lives are summarized in the timelines of Figure 7.3.

We will see that, for these five, solving equations was a passionate undertaking. Researchers today may also be passionate about their work, but academic conventions regarding publication and priority have tempered the overt expression of strong negative feelings.

Scipione del Ferro became a lecturer in arithmetic and geometry at the University of Bologna in 1496. It is not known exactly how or when del Ferro became interested in cubic equations. He discovered a remarkable new method for solving these equations, and then he did what no researcher today would do — he tried to keep his results secret. There is no way to know if del Ferro made his discoveries independently or if there are others who should share the credit. There are no surviving writings from him, although it is said that he wrote of his discoveries in a notebook that on his death passed to his son-in-law, Hannibal Nave, who was also a mathematician. Del Ferro, shortly before his death, communicated his method for solving cubic equations to his student, Fior. A mediocre mathematician Fior tried to use this information to elevate his reputation. Hoping to gain fame from his knowledge of del Ferro's secret, Fior challenged another mathematician, Tartaglia, to a public competition in 1535. However, when the competition took place on 13 February, it appeared that Tartaglia also knew how to solve cubic equations. In fact, Tartaglia had the greater skill and won a clear victory.

Niccolò Fontana (Tartaglia), still a child in 1512, was nearly killed by a French soldier during the sack of Brescia. He received a saber cut to his

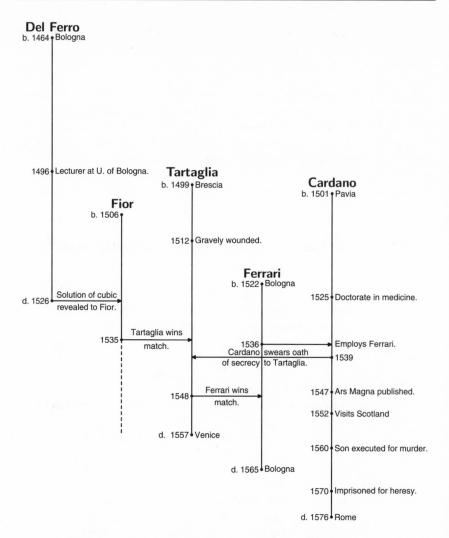

Figure 7.3. Five sixteenth-century Italian mathematicians.

jaw and palate, causing a speech impediment that occasioned his nick-name *Tartaglia*, "the stammerer." Tartaglia taught himself mathematics and earned a living teaching the subject in Venice and Verona. He acquired a reputation as a talented mathematician, and Fior considered him a worthy opponent. Tartaglia achieved further recognition because of his victory in the public contest with Fior. That Tartaglia was able solve cubic equations came to the notice of Girolamo Cardano, a mathematician who had attempted to solve these equations without success.

Thus, methods for solving cubic equations were discovered by del Ferro and rediscovered by Tartaglia. Nevertheless, this technique is often called *Cardan's method* after Girolamo Cardano. We will see how the credit for this discovery unfairly passed to Cardano.

Girolamo Cardano, illegitimate child of Fazio Cardano and Chiara Micheria, was born in Pavia in 1501. His mother had fled to that city from Milan to avoid an outbreak of the plague. Cardano learned mathematics from his father. He studied medicine at Pavia and Padua and received a doctorate in 1525. Cardano's life was hindered by two faults: he was quarrelsome — this made it difficult for him to obtain and hold academic positions — and he was a compulsive gambler.

In 1536, Cardano employed as a servant a 14-year-old orphan named Ludovico Ferrari. Cardano discovered that this boy was exceptionally intelligent and decided to teach him mathematics. Ferrari became a brilliant mathematician and a defender of Cardano against his enemies. In 1540, Cardano resigned his academic post at the Piatti Foundation in Milan so that his place could be taken by the 18-year-old Ferrari, who now became a public lecturer on geometry.

In 1539, Cardano wheedled from Tartaglia his secret method for solving cubic equations. Cardano swore an oath that he would not divulge the method before Tartaglia did so.

Four years later, Cardano and Ferrari traveled to Bologna and saw the notebook of del Ferro that was now in the possession of his son-in-law Hannibal Nave. Cardano now felt that he was released from his oath to Tartaglia because the method of solving cubic equations was discovered by del Ferro, not Tartaglia. Cardano reasoned that his oath did not prevent him from publishing del Ferro's method, even if it was largely identical to Tartaglia's.

In 1545, Cardano published his masterpiece, *Ars Magna* ("the great art"), which contains, among other things, the method of del Ferro–Tartaglia for solving cubic equations. Although Cardano gave credit to del Ferro and Tartaglia, posterity gave the credit to Cardano, perhaps because he was the first to publish. *Ars Magna* also contains Ferrari's solution of quartic equations — equations of the fourth degree like $x^4 - x + 1 = 0$.

Tartaglia formed an intense hatred of Cardano because Tartaglia felt that Cardano broke his oath of secrecy. Ferrari quickly backed his master in this quarrel and began an exchange of bitter insults with Tartaglia. Ferrari challenged Tartaglia to a public mathematical contest, but Tartaglia was reluctant to risk his reputation against the relatively unknown Ferrari. He would have preferred a match against Cardano, who had become a famous mathematician from the publication of *Ars Magna*.

A problem-solving contest between Ferrari and Tartaglia finally took place on 10 August 1548. The match was won by Ferrari, who enjoyed an immediate rise to fame. Of various employment offers, he chose a position as a tax assessor.

We must give credit to these five mathematicians of the Italian Renaissance for bringing algebra to a new higher level even though we may smile at their braggadocio. No doubt each of them was motivated by higher motives than just the desire for acclaim: a love of learning and a genuine gift for mathematics. One might think that science would benefit if conflicts among researchers could be reduced or eliminated. On the contrary, science benefits from healthy competition. Fortunately, today, there is an important academic convention that harnesses the power of this rivalry — *open publication* with *anonymous peer review*.

In today's academic world, certain rules of openness govern the recognition of intellectual achievement — rules that did not exist in sixteenth-century Italy. Today, academic researchers gain recognition for their discoveries by making a public announcement — ideally, publication in a *refereed* journal. When an author sends an article to an editor of a refereed academic journal, the editor then forwards the article to an expert referee, who provides an opinion on the merit of the article. The referee must be a researcher in the same field as the author, a *peer* reviewer. The identity of the referee is known only to the editor — not to the author or the public.

The five Italian mathematicians wrangled over ideas that had no commercial application: the solution of cubic equations — that is, equations of the third degree, like $x^3 - 6x^2 + 11x - 6 = 0$. For an idea with commercial application, patent laws are now available. However, even today, it is difficult to gain proper recognition for a commercial mathematical innovation because mathematical algorithms are not patentable.

These five mathematicians of the Italian Renaissance treated the method of solving cubic equations as a gem of great value. Despite their efforts to steal the gem and hide it from sight, it is now on view for all to see. We will now see some of this gem's mathematical facets.

Solving a cubic equation

We illustrate the general methods for solving cubic equations with one example:

$$x^3 + 6x + 2 = 0 \qquad (7.15)$$

Instead of the method given by Cardano in *Ars Magna*, we will use a method developed later.

The solution requires some remarkable insights that came only after much experimentation and many false starts. We will use a generalization of the good/bad method described above for solving quadratic equations. However, we need three categories of numbers instead of two. Let's call them "good," "bad," and "ugly." As before, rational operations applied to the coefficients of the equation give us the good numbers. Bad and ugly numbers involve cube roots. The numbers $\sqrt[3]{2}$ and $1/\sqrt[3]{2}$ are prototypes for bad and ugly, respectively, because they illustrate the following properties:

1. The product of a bad number multiplied by an ugly one is good.

$$\sqrt[3]{2} \text{ (bad) } \times \frac{1}{\sqrt[3]{2}} \text{ (ugly) } = 1 \text{ (good)}$$

2. The cube of a bad or ugly number is good.

$$\left(\sqrt[3]{2} \text{ (bad) } \right)^3 = 2 \text{ (good)}$$

$$\left(\frac{1}{\sqrt[3]{2}} \text{ (ugly) } \right)^3 = \tfrac{1}{2} \text{ (good)}$$

We assume that the good, bad, and ugly quantities satisfy the multiplication table in Table 7.2.

After a hundred false starts, the hundred-and-first is the following brilliant guess: *Maybe the solution can be expressed as the sum of a bad quantity v and an ugly one w.* Let us see where this leads.

Making the substitution $x = v + w$, we use the identity

Table 7.2. Good, bad, and ugly multiplication table.

	good	bad	ugly
good	good	bad	ugly
bad	bad	ugly	good
ugly	ugly	good	bad

$$(v + w)^3 \equiv v^3 + 3v^2 w + 3vw^2 + w^3$$

Equation (7.15) becomes

$$v^3 + 3v^2 vw + 3vw^2 + w^3 + 6v + 6w + 2 = 0$$

It follows from Table 7.2 that the terms u^3 and v^3 are good. Including also the constant 2, the good terms are

$$u^3 + v^3 + 2$$

The remaining terms are

$$3v^2 w + 3vw^2 + 6v + 6w \equiv v(3vw + 6) + w(3vw + 6)$$

We can group the good, bad, and ugly terms of equation (7.15) as follows:

$$\underbrace{v^3 + w^3 + 2}_{\text{good}} + \underbrace{v(3vw + 6)}_{\text{bad}} + \underbrace{w(3vw + 6)}_{\text{ugly}} = 0 \qquad (7.16)$$

Because we expect an incompatibility between good, bad, and ugly, we seek to make each of these three parts separately equal to zero. We see that both the bad and ugly parts are zero if we choose v and w such that

$$vw = -2 \qquad (7.17)$$

Additionally, to make the good part equal to zero, we must have

$$v^3 + w^3 + 2 = 0 \qquad (7.18)$$

Consider now that v and w are fixed solutions of equations (7.17) and (7.18). We now write an equation that is quadratic with respect to a new variable z and has roots v^3 and w^3:

$$(z - v^3)(z - w^3) \equiv z^2 - (v^3 + w^3)z + v^3 w^3 = 0 \qquad (7.19)$$

We use equations (7.17) and (7.18) to simplify equation (7.19):

$$z^2 + 2z - 8 = 0 \qquad (7.20)$$

The roots of equation (7.20) are $z = -4$ and $z = 2$. (This can be seen, e.g., by applying the quadratic formula, equation (7.14).)

On the other hand, the roots of equation (7.20) are v^3 and w^3, and $v + w$ is a root of the original cubic equation (7.15). In other words,

$$x = -\sqrt[3]{4} + \sqrt[3]{2} = 0.32748\ldots$$

is one of the roots of the cubic equation (7.15). We can confirm this fact by substituting $x = 0.32748$ in the left side of equation (7.15).

Comments on the above solution:

1. Additionally, it can be shown that there are two complex[2] solutions of equation (7.15).

2. The technique that was applied above to the specific equation (7.15) can also be applied to any cubic equation of the form

$$x^3 + px + q = 0 \qquad (7.21)$$

Furthermore, *any* cubic equation can be reduced to this form by means of a suitable transformation.

3. There is an additional complication in the solution of equation (7.21) if the quadratic equation that takes the place of equation (7.20) has complex roots. In this case, it can be shown that the cubic equation (7.21) has three real roots. It is curious that an excursion into the complex numbers is needed in order to obtain real solutions.

4. Cube roots arise in the solution of cubic equations. In general, the solution also involves square roots that arise from the solution of a quadratic equation like equation (7.20).

5. The above method is diabolically clever, and it is elegant in that it shows that a solution can be found using square and cube roots. However, if one needs a numerical solution of a cubic equation, there are easier iterative methods, like the Babylonian square root algorithm illustrated in

Figure 6.6. Also note that the above method does not achieve a numerical solution. After finding $x = -\sqrt[3]{4} + \sqrt[3]{2}$, we still must find numerical approximations for the cube roots.

After the solution of cubic and quartic equations, it would seem that the fifth and higher degree equations might be next to fall. This never happened, and it never will happen because in 1824 the Norwegian mathematician Niels Henrik Abel (1802–29) proved that it is impossible to solve the fifth degree equation in the same sense that del Ferro and Tartaglia solved the cubic and Ferrari solved the quartic. This impossibility was independently rediscovered in 1829 by Évariste Galois (1811–32). Galois's methods have a significance that goes far beyond this particular result, methods that belong to a field of mathematics now known as *Galois theory*.

The Italian mathematicians of the sixteenth century did not solve equations in order to meet the needs of their technology. Today's technology needs the solution of all kind of equations, which are generally solved by numerical methods instead of the techniques of Cardano and the others.

An important motive for these sixteenth-century Italian mathematicians was to show who possessed the greatest mathematical power. However, we should not write off their efforts as mere self-aggrandizement. In fact, their solution of equations using radicals (square and cube roots, etc.) two centuries later led Galois and others to discover the theory of groups, a fundamental concept today in atomic physics and elsewhere in science and technology. In the next chapter, we will see that group theory also leads to an understanding of the ornamentation of the Alhambra.

8

Symmetry Without Fear

Tyger Tyger, burning bright,
In the forests of the night;
What immortal hand or eye,
Could frame thy fearful symmetry?

—WILLIAM BLAKE (1757–1827), Songs of Experience

SYMMETRY IS ORGANIZED REPETITION, seen both in the tiger's stripes and in Blake's poetry; also, in the tile work of the Alhambra,[1] the fugues of Bach, the biology of the starfish, the growth of crystals, the shape of galaxies, and the theory of subatomic particles. The *bilateral symmetry* of the human body is only the first page of the intricate book of symmetry.

In art, literature, and music, symmetry is a framework for artistic expression—the backstage scaffoldings, ropes, and pulleys. Artistic symmetry is often more effective if the principle of repetition is not immediately evident.

On the other hand, mathematics wants to see the bare bones, the essence, of symmetry. We will see that symmetry has a connection with algebra—more specifically, *abstract* algebra. We will see that symmetries are geometric transformations with precise mathematical definitions, and that symmetries are endowed with algebraic operations that resemble numerical multiplication.

Symmetric ornaments of the two-dimensional plane exhibit the connection between art and mathematics.[2] The creation of beautiful symmetric plane ornaments requires artistic skill combined with at least an implicit understanding of certain mathematical relationships. In this chapter, we are concerned with plane ornaments that are repetitive in the sense that they can be brought into coincidence with themselves by suitable motions.

142

Symmetric plane ornaments are divided into three categories: *borders* (e.g., Figure 8.1), *wallpaper designs* (e.g., Figure 8.2), and *rosettes* (e.g., Figure 8.4). A *border* is a decorative horizontal band often found on the upper part of a wall. The term *wallpaper design* refers to designs found not only on paper wall coverings but also on fabrics and ceramic tile work.

More complex designs arise (1) by increasing the dimension from two to three or (2) by introducing color as an added element of symmetry. Crystallography is the study of three-dimensional patterns. In this chapter, we examine two-dimensional monochromatic ornaments only; they are sufficient to illustrate the richness of the connection with algebra.

In the real world, border and wallpaper designs are finite. However, for the purpose of mathematical analysis, we adopt the fiction that these designs have infinite extent.

In this chapter, we make a careful distinction between *design* and *pattern*. A design is a particular graphic representation, but a pattern is a scheme that underlies many designs. Although there are infinitely many different border and wallpaper designs, there are only seven border *patterns* and 17 wallpaper *patterns*.

Each pattern is governed by a mathematical structure called a *group*. Each different symmetry pattern has a different group. The next section illustrates the connection between plane ornaments and algebra by discussing in detail a particular group. It is hoped that this example has enough structure to be interesting and yet is simple enough to show the basic concepts.

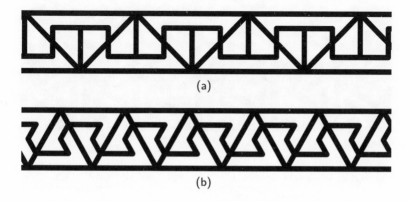

(a)

(b)

Figure 8.1. Border ornaments, from a Persian manuscript.[3]

(a)

(b)

Figure 8.2. Wallpaper designs, based on wall paneling from the Alhambra.[4]

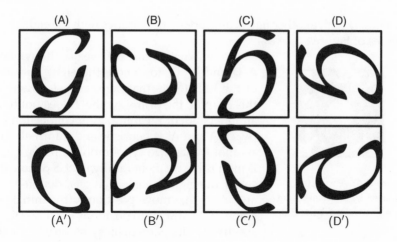

Figure 8.3. The eight symmetric images of a reversible square frame.

Symmetries of a Square

A square is a rosette of austere simplicity. The symmetries of a square illustrate the connection between symmetry and algebra. Until now, we have been using the word *symmetry* loosely, but, for this discussion, we will use a more precise definition. *A symmetry of the square is a motion that brings the square into coincidence with itself.*

The vacuous motion that moves nothing, called the *identity*, counts as one of the symmetries. How many more there are depends on what motions are allowed. If we are only allowed to rotate the square by sliding it in its initial plane, then there are four symmetric images of the plane — rotations by a multiple of 90° — but if we are allowed to turn over the square, then there are four more, making a total of eight.

Figure 8.3 shows the symmetries of a square that can be rotated or reflected; we call it a *reversible square frame* (RSF). Think of an RSF as an empty square picture frame that looks exactly the same from both sides. The symmetries (including the identity) of an RSF are eight in number; they correspond with the eight different ways of placing a slide in a projector. These are eight motions — rotations possibly combined with reflection — that bring the RSF back into coincidence with itself. Application of the eight symmetries to an RSF results in eight indistinguishable blank squares. In order to distinguish the symmetries, Figure 8.3 shows the letter G inside the square.

The RSF is not the only plane figure that exhibits the symmetries shown in Figure 8.3. For example, the rosette in Figure 8.4 is obtained by superimposing the eight images in Figure 8.3 of the letter G.In fact, substituting the letter $\mathcal{G}$ with another figure produces other rosettes with the eight symmetries of Figure 8.3.

Figure 8.4. A rosette obtained by superimposing the eight $\mathcal{G}$s in Figure 8.3.

As shown in Figure 8.3, we use the symbols A–D and A$'$–D$'$ to denote the eight positions of the RSF. (Note that here, and elsewhere in this chapter, the prime symbol $'$ does not have an independent meaning. Any connection between A and A$'$ is heuristic and informal. The symbol A$'$ has two typographical parts, but it is indivisible mathematically.) Let S denote the set of the eight transformations, denoted a–d and a'–d', that move position A (the initial position of the RSF) into positions A–D, A$'$–D$'$, respectively. In particular, symmetry b rotates position A a quarter turn (90°) counterclockwise into position B. Moreover, symmetry b rotates *any* of the eight positions in the same manner:

$$b : A \rightarrow B, \ B \rightarrow C, \ C \rightarrow D, \ D \rightarrow A, \ A' \rightarrow D', \ B' \rightarrow A', \ C' \rightarrow B', \ D' \rightarrow C'$$

The position that results from applying symmetry b to position C is denoted bC, so that we have bC $=$ D.

The symmetry a is the identity transformation because it leaves the square unmoved, for example, aC $=$ C. The other symmetries all have more than one geometric description; for example, b can be considered either a 90° counterclockwise rotation or a 270° clockwise rotation, and a' is either a reflection through a horizontal line or a reflection through a vertical line followed by a 180° rotation.

We define *multiplication of symmetries* as follows: The application of symmetry b to the result of applying c to D is denoted $b(c$D$)$, or, without ambiguity, we omit the parentheses and write bcD. Carrying out this computation, we have bcD $= b$A $=$ B. We say that the product bc applied to D gives B. Since c is a 180° rotation and b is a 90° counterclockwise rotation, bc is a 270° counterclockwise rotation—the same as a 90° clockwise rotation. Therefore, bc is the same symmetry as d. Multiplication of any two symmetries in S is defined similarly. Table 8.1 is the multiplication table for all eight symmetries in S.

The group, also called the symmetry group, *of the RSF is the set of these eight symmetries together with the multiplication defined in Table 8.1.* More generally, a group is set of elements—not necessarily symmetries—together with a definition of multiplication of pairs of elements satisfying the three axioms listed on page 148.

We will discuss the many ways in which the multiplication in Table 8.1 resembles ordinary multiplication, but it fails to do so in one respect. For ordinary multiplication, the order of the factors does not affect the product; for example, $3 \times 4 = 4 \times 3$. This is an instance of the *commutative law*

Table 8.1. Multiplication table for the symmetries $\mathcal{S}$ of a reversible square frame. The bold entries indicate instances of noncommutativity, for example, $a'b = b'$, $ba' = d'$.

		Pure rotation				Reflection & rotation			
		a	b	c	d	a'	b'	c'	d'
Pure rotation	a	a	b	c	d	a'	b'	c'	h'
	b	b	c	d	a	**d'**	**a'**	f	c'
	c	c	d	a	b	c'	h'	a'	b'
	d	d	a	b	c	f	c'	d'	a'
Reflection & rotation	a'	a'	**f**	c'	**d'**	a	b	c	d
	b'	b'	c'	h'	a'	d	a	b	c
	c'	c'	d'	a'	f	c	d	a	b
	h'	h'	a'	b'	c'	**b**	c	d	a

of multiplication: for any numbers x and y, we have $xy = yx$. There are 12 instances in Table 8.1 in which the commutative law fails. Figure 8.5 shows one instance of this: $a'b = b'$ and $ba' = d'$.

The symmetries a', b', c', d' have the following geometric meanings:

a': A reflection across a horizontal line.

b: A 90° rotation counterclockwise.

b': A reflection across a horizontal line followed by a 90° rotation counterclockwise.

d': A reflection across a horizontal line followed by a 90° rotation clockwise.

$a'b = b'$

$ba' = d'$

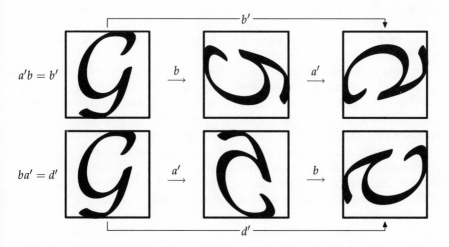

Figure 8.5. Symmetries are noncommutative: $a'b \neq ba'$. The top and bottom rows illustrate, respectively, that $a'b = b'$ and $ba' = d'$.

If X is any of the positions A–D and A′–D′, $a'b$X means the same as $a'(b$X$)$. In other words, in the product $a'b$ the symmetry b must be applied first. Thus, the product $a'b$ denotes a rotation 90° counterclockwise followed by a reflection across a horizontal line. On the other hand, the product ba' denotes reflection across a horizontal line followed by a rotation 90° counterclockwise.

The Group Axioms

Unlike the ordinary multiplication of numbers, the multiplication of symmetries of a square is noncommutative—as shown in the preceding section. Nevertheless, the multiplication of symmetries shares several properties with ordinary multiplication. These properties are called *axioms*. Any set G of elements with an operation of multiplication that satisfies the following three axioms is called a *group*—a name coined in 1830 by Évariste Galois (1811–1832).[5] For example, S, defined in the preceding section, is a group.

Axiom 1. The associative law of multiplication. *A product of three factors gives the same result whether we start by multiplying the first two factors or the last two. In other words, for any three elements x, y, and z of G, we have* $(xy)z = x(yz)$. This familiar property of the multiplication of numbers is true for the multiplication of symmetries in S because, for any position W, $(xy)z$W and $x(yz)$W both have the same meaning: "first transform W with symmetry z, then y, and, finally, x."

Axiom 2. The identity. *There exists an element e of S, called the identity, such that, for any element x of G, we have ex $= xe = x$.* For example, the number 1 has this property: for any number x, we have $1 \cdot x = x \cdot 1 = x$. Similarly, the identity symmetry a has the property that for any symmetry x in S, we have $ax = xa = x$.

Axiom 3. The inverse. *For any element x of G, there exists exactly one element x^{-1} of G, called the inverse of x, such that $x^{-1}x = xx^{-1} = e$ where e is the identity for G.* For example, for any nonzero number x, there is exactly one number y, called the reciprocal of x, such that $xy = 1$. Similarly, for any symmetry x, there exists exactly one symmetry y such that xy is equal to the identity, that is, such that $xy = a$. For example, since b is a 90° counterclockwise rotation, the left inverse of b is d because d is 90° *clockwise* rotation: $bd = db = a$. The rotation d negates the rotation b.

Since the multiplication of symmetries is noncommutative, it is conceivable (but we will see that it is impossible) that there might be a pair of symmetries, x and y, such that y is a right, but not a left, inverse of x. In other words, we might have $xy = a$ and $yx \neq a$. However, Axiom 3 asserts that this cannot happen—that every element x of G commutes with its inverse $(x^{-1}x = xx^{-1})$. This can be confirmed for S by examining Table 8.1.

This table confirms the fact, noted above, that the symmetries b and d are inverse to each other ($b^{-1} = d$) because the table shows $bd = db = a$. (Recall that a is the identity symmetry.) The six other symmetries of S are their own inverses[6] because:

$$aa = cc = a'a' = b'b' = c'c' = d'd' = a$$

Axioms 2 and 3 are slightly stronger than necessary. It can be shown that if merely a *right* identity e exists, then e is also a *left* identity. In other words, if $xe = x$ for all x in G, then it is also true that ex is equal to x for all x in G. Similarly, if merely a *right* inverse of x exists, that is, if there exists y in G such that $xy = e$, then y is also a *left* inverse of x: $yx = e$. Hence, we write $y = x^{-1}$, and it is not necessary to use separate notations for left and right inverses.

An examination of Table 8.1 shows that S is a group. However, this table is a redundant description of S. In fact, there is a simpler way of describing the symmetries of a square that comes closer to the way in which one rotates and/or reverses a transparency in a projector to achieve the images in Figure 8.3. One reflection and one 90° rotation suffice to generate all the needed transformations. More specifically, repeated application of the two symmetries b (the 90° counterclockwise rotation) and a' (the horizontal reflection) generates all of the symmetries of the group S, apart from the identity. For this reason, these two elements of S, b and a', are called *generators* of S. For example, the symmetry c' is the same as two counterclockwise 90° rotations (b^2) followed by a horizontal reflection (a'). This is expressed by the formula $c' = a'b^2$. (Note that, in this formula, the rightmost symmetry, b, is applied first.) The following is a complete list showing representations of the nonidentical symmetries in terms of a' and b:

$$c = b^2 \qquad\qquad d = b^3$$
$$b' = a'b \qquad c' = a'c = a'b^2 \qquad d' = a'd = a'b^3$$

The symmetries b and a' are not the only generators of S. A *clockwise* 90° rotation (d) and *vertical* reflection are also generators. As remarked earlier, d is a clockwise 90° rotation. Examination of Figure 8.3 shows that c' is a vertical reflection. Thus, d and c' are an alternate choice for generators of the group S.

All the rotations and reflections of group S leave a certain point fixed: the center of the square. We have seen that the square itself and Figure 8.4 are transformed into themselves by all the symmetries of S. Figures with this invariance property are called rosettes because many have decorative uses. There are other groups of rotations and reflections that leave one point fixed — for example, the group of rotations by a multiple of 120°. Each such group has an associated family of rosettes.

The following are a few examples of groups:

1. The set of all integers (positive, negative, and zero) is a group under addition. This means that the group multiplication is understood to be ordinary addition. For this group, the identity is the number 0, and the inverse of a number a is $-a$.
2. The set of nonzero real numbers under ordinary multiplication: The identity is 1, and the inverse of a is $1/a$. The number 0 must be excluded because it fails to have an inverse. There are many subsets of the real numbers that are also groups under ordinary multiplication:
   (a) The positive rational numbers
   (b) Positive numbers of the form $a + b\sqrt{2}$ where a and b are rational
3. The numbers 0, 1, ..., 9 where the product of two numbers is defined to be the remainder on dividing the ordinary sum by 10. For example, the group product of 7 times 8 is equal to 5 because the remainder on dividing $7 + 8 = 15$ by 10 is equal to 5.
4. The numbers 1, 2, ..., 10 where the product of two numbers is defined to be the remainder on dividing the ordinary product by 11. The proof that this is a group makes use of the fact that 11 is a prime. The number 1 is the identity. The group product of 7 times 8 is equal to 1 because the remainder on dividing $7 \times 8 = 56$ by 11 is 1. This calculation also shows that 7 and 8 are inverse to each other in this group.

Isometries of the Plane

Rotations and reflections are examples of isometries of the plane. An *isometry* is a transformation of the plane onto itself that leaves all distances unchanged. The set of isometries of the plane satisfies Axioms 1–3 and, therefore, constitutes a group. An isometry can be visualized by repositioning a piece of paper on a flat desk. Instead of a piece of paper, a transparency (of the sort used in an overhead projector) is a more apt illustration because the image on the transparency is still visible when the sheet is turned upside down. Of course, we ignore the fact that the sheet is finite whereas the concept of a plane is infinite.

In addition to rotation and reflection, there is a third type of isometry: translation. The three isometries are visualized by the following movements of the transparency sheet:

- *Rotation.* Move the sheet so that one point remains fixed. For example, spin the sheet with one point fastened to the desk with a pin. A rotation is specified by giving the fixed point and a clockwise or counterclockwise angle.
- *Reflection.* Turn the sheet upside down in such a way that all points on a certain line, the "mirror," keep their original positions.

- *Translation.* Slide the sheet so that the final position of each line is parallel to its original position. A translation is specified by two distances: horizontal movement right or left and vertical movement up or down.

In addition to these three fundamental symmetries, there is a fourth, *glide reflection* that is important in border and wallpaper symmetry groups. Glide reflection is a composite of reflection and translation. This symmetry is illustrated in Figure 8.6. In this figure, the dart *a* is moved to the position of dart *b* by subjecting the entire design to the following two transformations:

1. Advance the entire design horizontally 0.2 inch.
2. Then reflect the design across the horizontal line shown in Figure 8.6.

This motion is an example of a *glide reflection.* In general, a glide reflection is a reflection followed by a translation in the direction of the line of reflection.

The full isometry group $\mathcal{I}$ consists of all rotations, reflections, and translations of the two-dimensional plane. The group $\mathcal{S}$, discussed above, is a *subgroup* of $\mathcal{I}$. This means that (1) the group $\mathcal{S}$ is a subset of $\mathcal{I}$, and (2) the group multiplication in $\mathcal{S}$ is inherited from $\mathcal{I}$.

Each of the patterns for two-dimensional ornaments—rosettes, borders, and wallpaper designs—generates a distinct subgroup of $\mathcal{I}$.

Patterns for Plane Ornaments

Patterns for rosettes are simple. The rosette group for the reversible square frame (discussed above) is generated by a quarter turn and a reflection through the center of rotation. In general, a rosette group (or pattern) is generated by a $1/n$ turn (where n is a natural number), with or without a reflection through the center of rotation. Thus, for each natural number n, there are two different rosette groups.

Catalog of border and wallpaper patterns

Figures 8.7 and 8.8 catalog the seven border patterns and the 17 wallpaper patterns, respectively. For each pattern, that is, for each symmetry group,

Figure 8.6. A border design.

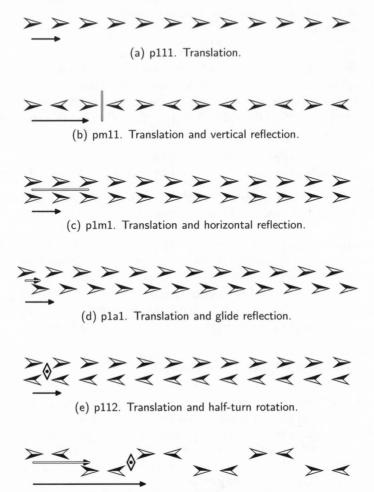

(a) p111. Translation.

(b) pm11. Translation and vertical reflection.

(c) p1m1. Translation and horizontal reflection.

(d) p1a1. Translation and glide reflection.

(e) p112. Translation and half-turn rotation.

(f) pma2. Translation, glide reflection, and half-turn rotation.

(g) pmm2. Translation, horizontal and vertical reflections, and half-turn rotation.

Figure 8.7. The seven border patterns.

⟶ Translation ═══ Reflection ═══▷ Glide reflection ◊ Half turn

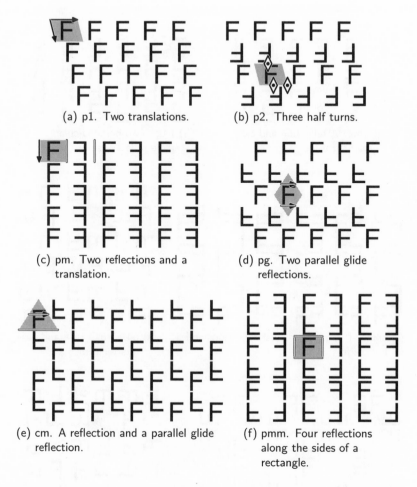

(a) p1. Two translations.

(b) p2. Three half turns.

(c) pm. Two reflections and a translation.

(d) pg. Two parallel glide reflections.

(e) cm. A reflection and a parallel glide reflection.

(f) pmm. Four reflections along the sides of a rectangle.

Figure 8.8. The 17 symmetries of the plane. (a)–(q) The shaded regions are the basic repeated units. Each wallpaper pattern is generated by repeated application of the translations, reflections, glide reflections, and rotations indicated by the following symbols:

		Rotations:	
⟶	Translation	◈	Half turn
=	Reflection	△	120° turn
⇒	Glide reflection	▣	Quarter turn

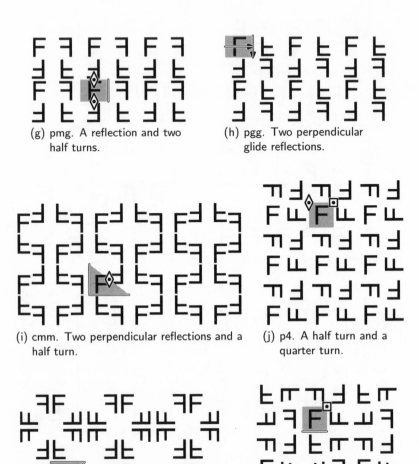

(g) pmg. A reflection and two
    half turns.

(h) pgg. Two perpendicular
    glide reflections.

(i) cmm. Two perpendicular reflections and a
    half turn.

(j) p4. A half turn and a
    quarter turn.

(k) p4m. Reflections in three sides of
    an isosceles right triangle.

(l) p4g. A reflection and a
    quarter turn.

Figure 8.8. Continued.

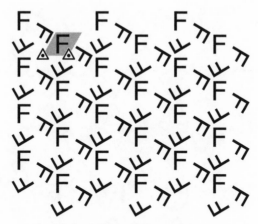

(m) p3. Two 120° rotations.

(n) p31m. A reflection and a 120° rotation.

Figure 8.8. Continued.

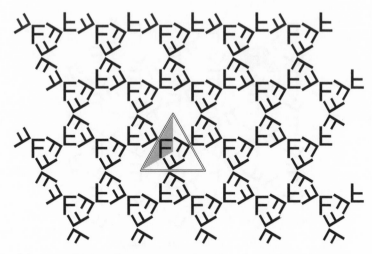

(o) p3m1. Three reflections along the sides of a equilateral triangle.

(p) p6. A half turn and a 120° turn.

Figure 8.8. Continued.

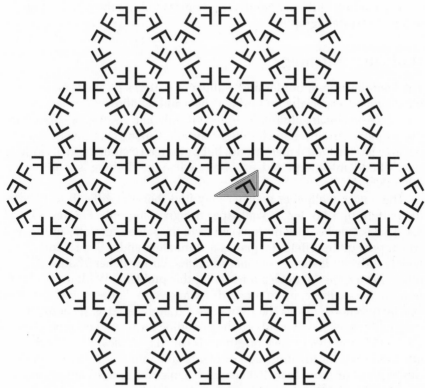

(q) p6m. Three reflections along the sides of a 30°–60° right triangle.

Figure 8.8. Continued.

the characteristic symmetries are listed and shown in the figures by special symbols for translations, reflections, glide reflections, and rotations. In each case, these symmetries generate a different symmetry group.

Although there is a limited number of these patterns, there is an unlimited number of designs based on them. The finite number of patterns does not limit artistic expression.

Each symmetric design in Figures 8.7 and 8.8 is generated by rotations, reflections, and translations of a fundamental figure. For the border patterns (Figure 8.7) the fundamental figure is the dart ➤, and for the wallpaper patterns (Figure 8.8) it is the letter F. The requirement for a fundamental figure like ➤ or F is that it should be *asymmetric* in order not to introduce spurious symmetries. In Figure 8.8, one instance of the fundamental figure is shown within a shaded region; the pattern symmetries transform each shaded region into copies of itself that cover the entire plane without overlap.

The naming of these patterns (e.g., pm11) follows the standard terminology of crystallography.[7]

Wallpaper watching

The identification of border and wallpaper patterns provides an interesting diversion, especially for bird watchers and wildflower fanciers during the off season. Such patterns are found on walls and elsewhere — for example, carpets, curtains, and clothing. Caution: Gazing excessively at a carpet may be considered merely odd, but prolonged examination of clothing can be badly misinterpreted, especially when the clothing is worn by another person.

The novice wallpaper watcher may find that it is difficult to identify the pattern of a border or wallpaper design using only Figures 8.7 and 8.8. This is true because designs can have the same symmetry group and still appear quite different — just as a chihuahua and a Saint Bernard are widely different instances of *Canis familiaris*. Bird watchers, botanists, and others have a special device for solving this problem. Wildflower identification, for example, often proceeds by answering a prescribed series of yes–no questions: "Does the stem have a triangular cross-section?" Each succeeding question depends on the preceding answer — yes or no. An entire hierarchical tree of such questions is called a *taxonomic key*. Such keys are often used for identification of plants and animals. A taxonomic key can be short or more than 1000 pages.[8] Tables 8.2 and 8.3 are taxonomic keys for the identification of border and wallpaper patterns. The reader may want to use these keys to identify the patterns of Figures 8.1 and 8.2.[9]

Table 8.2. Key to the seven border patterns. Based on Washburn and Crowe (1988, Table 4.1, p. 83), with permission from University of Washington Press.

Start: Is there a vertical mirror line?
   Yes: Is there a horizontal mirror line?
      Yes: Figure 8.7(g) *pmm2*
      No: Is there a half turn?
         Yes: Figure 8.7(f) *pma2*
         No: Figure 8.7(b) *pm11*
   No: Is there a horizontal mirror line or a glide reflection?
      Yes: Is there a horizontal mirror line?
         Yes: Figure 8.7(c) *p1m1*
         No: Figure 8.7(d) *p1a1*
      No: Is there a half-turn?
         Yes: Figure 8.7(e) *p112*
         No: Figure 8.7(a) *p111*

Table 8.3. Key to the 17 wallpaper symmetries — from Reid (1999). Based on Washburn and Crowe (1988, Table 5.1, p. 128), with permission from University of Washington Press.

Start: Does the pattern reflect in at least one direction?
Yes: Does the pattern rotate?
    Yes: What is the smallest rotation?
    180°: Are there reflections in two directions?
        Yes: Are all the centers of rotation on mirror lines?
            Yes: Figure 8.8(f) .... *pmm*
            No: Figure 8.8(i) .... *cmm*
        No: Figure 8.8(g) .... *pmg*
    120°: Are all the centers of rotation on mirror lines?
        Yes: Figure 8.8(o) *p3m1*
        No: Figure 8.8(n) *p31m*
    90°: Are there mirror lines at 45°?
        Yes: Figure 8.8(k) *p4m*
        No: Figure 8.8(l) *p4g*
    60°: Figure 8.8(q) *p6m*
    No: Is there a glide reflection which doesn't lie on a mirror line?
    Yes: Figure 8.8(e) .... *cm*
    No: Figure 8.8(c) .... *pm*
No: Does the pattern rotate?
    Yes: What is the smallest rotation?
    180°: Is there a glide reflection?
        Yes: Figure 8.8(h) .... *pgg*
        No: Figure 8.8(b) *p2*
    120°: Figure 8.8(m) *p3*
    90°: Figure 8.8(j) *p4*
    60°: Figure 8.8(p) *p6*
    No: Is there a glide reflection?
    Yes: Figure 8.8(d) .... *pg*
    No: Figure 8.8(a) *p1*

Since its discovery by Galois in 1830, the concept of group has been central to the study of mathematics beyond the elementary level. Galois used groups to study the roots of algebraic equations. In this chapter, we introduced the symmetry groups of two-dimensional ornaments. This is only the beginning. In modern mathematics, groups are everywhere.

We have seen in this chapter that the study of abstract algebra goes far beyond the treatment of arithmetic problems characteristic of elementary algebra. In the next chapter, we will extend the realm of abstract algebra in a way that might seem bizarre. We will see that algebra can examine itself and, indeed, *all of mathematics.*

9

The Magic Mirror

Undecidability

A MAGIC MIRROR that answers every question truthfully would be a scientist's dream. Common sense suggests that this is only a fairy tale and no such truth-machine can exist. In fact, if there were such a mirror, we could confound it with the following question:

— Is the following a true statement? **I am lying.**

A little reflection shows that neither *Yes* nor *No* can be a truthful answer. If the mirror says *Yes*, then it should have said *No*; and if it says *No*, then it should have said *Yes*. The best that the mirror can do is to say:

— That's a smart-aleck question. Ask me a *real* question.

The above is a version of the *liar's paradox*—first stated by the Cretan Epimenides in the seventh century BCE. Epimenides said, "All Cretans are liars." Did he lie or tell the truth?

One might think that the search for a truth machine died out with the practice of witchcraft and alchemy. Such was not the case. Early in the twentieth century, the search for a truth machine was a serious, respectable, and plausible concern of mathematical logic. This quest was supported by a movement called *formalism* — led by David Hilbert (1862–1943), a very distinguished and influential German mathematician of the University of Göttingen. Of course, the formalists were too sophisticated to expect to literally find a magic mirror. They narrowed their concern to *mathematical* questions that can be answered *Yes* or *No*. The formalists thought that it might be possible to determine the truth of a mathematical assertion in two steps:

1. Translate the assertion into the formal language of symbolic logic. Sometimes this might be tedious, but no one denies that it can always be done.
2. Find an algorithm (a prescribed step-by-step calculation) called a *decision procedure* to determine the truth or falsity of any such formally stated assertion. To carry out the procedure one would merely follow certain prescribed rules without considering the meaning of the assertion. This second step is the where the difficulty lies. No one has overcome this difficulty for the very good reason that there is a mathematical proof that this task is impossible.

The existence of a decision procedure was conjectured in 1928 in a book by Hilbert and Ackermann.[1] And it was disproved three years later in 1931 by Kurt Gödel (1906–78), who formulated and proved the *undecidability theorem*.[2] Gödel showed that any axiomatic system that is strong enough to include ordinary arithmetic is either *inconsistent* or *undecidable*.

Inconsistent means that it is possible to derive contradictions. It can be shown that if it is possible to derive any contradiction whatever, then it is possible to derive the particular contradiction $0 = 1$. No one seriously contends that ordinary arithmetic is inconsistent. Common sense suggests that ordinary arithmetic is consistent, but no one has *proved* that it is.

Undecidable means that the system permits meaningful arithmetic statements that can be neither proved nor disproved using only the axioms of the system.

Gödel proved his undecidability theorem by translating the smart-aleck liar's paradox above into an arithmetic proposition that denies its own truth. The details of Gödel's construction are quite complex, and we will not attempt to understand them. No doubt the formalists knew the liar's paradox, but they might not have seen its significance because it did not seem to be a *mathematical* question. Gödel's tour de force showed that, on the contrary, it is possible to transform the liar's paradox into an undecidable arithmetic proposition.

If a proposition is undecidable, it does not follow that its truth or falsity is inconclusive. One might think that if a proposition is undecidable then we cannot say whether it is true or false, but the following story shows that this is not always so.

The Tragic Case of the Decent Numbers

Ada and Ben have made a mathematical discovery, but, unfortunately they died in a fire that also consumed most of their unpublished manuscript. An examination of the ashes reveals only fragmentary information. We see that they define a class of the natural numbers that they call the *decent* numbers, but the details of that definition are illegible—except for the claim that the definition involves only ordinary logic and arithmetic. We can make out the following proposition.

Proposition 9.1. *All natural numbers are decent.*

The fragment also claims that the remainder of the manuscript—completely lost in the fire—consists of a proof that Proposition 9.1 is undecidable.

Let us assume that Ada and Ben correctly proved that Proposition 9.1 is undecidable. Does that mean that it is inconclusive whether Proposition 9.1 is true or false? No, Proposition 9.1 is certainly true. Indeed, if it were false then there must exist an *indecent* number, and we could *prove* that Proposition 9.1 is false merely by exhibiting this number—contrary to our assumption that Ada and Ben found a correct proof that Proposition 9.1 is undecidable. Proposition 9.1 is true, but it cannot be derived from the usual axioms of arithmetic using ordinary logic.

Gödel's discovery had a major effect upon mathematics and philosophy. It hastened the end of the optimism of the Enlightenment that began in the seventeenth century—the belief that rational thought has no limits and that the science of Newton and others would proceed without bound.[3] Gödel discovered a blemish in the axiomatic method, but this in no way diminishes its previous accomplishments dating back to the ancient Greeks. However, expectations of future successes rose too high. Gödel's theorem is like a minor stock market correction. We don't mind because we have invested for the long term.

The Magic Writing

Why did Hilbert and the formalists suspect that a decision method might exist? They were encouraged by a history of success going back hundreds of years. Mathematicians discovered a remarkable new way of writing mathematical assertions. To the novice, algebra may seem as unintelligible as is Chinese to one ignorant of that language. In fact, there are several mutually unintelligible Chinese languages that share a common written language. Algebra is similar to Chinese in this respect because algebra is an international written language.

Certain algebraic formulas (e.g., equations) are equivalent to sentences in ordinary language. The most remarkable feature of the language of algebra is that there is a body of rules that enables us to mechanically transform one algebraic assertion into another so that if the first assertion is true then the derived assertion is also true. Starting with the equation $x - 2 = 3$, we apply the rule that we may add the same number to both sides of an equation. Adding 2 to both sides, we obtain $x = 5$. In other words, if x is a number such that $x - 2 = 3$, then x must be equal to 5.

Although this basic pattern remains valid, the art of solving equations goes far beyond this trivial example. The scope of this method is enlarged by increasing the number of rules and the complexity of the initial assertions. The method is further extended applying algebraic methods to other areas of mathematics such as set theory and symbolic logic.

In elementary algebra, these abstract methods are applied to arithmetic assertions. The basic variables (x, y, etc.) represent (generally unspecified) numbers; formulas are formed using these variables together with arithmetic operations ($+$, $-$, $\times \div$), relations ($=$, $<$, $>$), and grouping symbols (e.g., parentheses and brackets). *Symbolic logic* augments arithmetic formulas by including further abbreviations that make it possible to state more complex assertions. In particular, symbolic logic includes abbreviations for *and* ($\wedge$), *or* ($\vee$), *not* ($\sim$), *for every* ($\forall$), *there exists* ($\exists$), *such that* (:), *implies* ($\Rightarrow$), and *if and only if* ($\Leftrightarrow$). Furthermore, symbolic logic permits variables (A, B, etc.) that represent (generally unspecified) assertions called *propositions*.

Example 9.1. (a) The expression

$$\sim (A \wedge B) \Leftrightarrow \sim A \vee \sim B$$

is read as

A and B are not both true if and only if either A is false or B is false.

(b) The expression

$$\forall x \, \exists y : y > x$$

is read as

For every number x there exists a number y such that y is greater than x.

Symbolic logic is a tool of *metamathematics*, the study of the methodology of proof in mathematics. A proof, First and always, must be a convincing argument. Although the manner in which mathematicians present proofs is quite varied, metamathematics assumes that all mathematical proofs can be expressed in symbolic logic. From this point of view, a proof, relative to a system of axioms, is a sequence of assertions, conforming to the rules of inference, that starts from the axioms, or a previously proven result, and terminates in a desired conclusion. In contrast, real proofs, as they are actually presented by mathematicians, use ordinary language, which is much less terse than symbolic logic; and real proofs use far fewer steps than symbolic logic would require.

It might seem that we have succeeded merely in creating an arcane code for writing mathematical assertions. Indeed, mathematicians are sometimes accused of creating a private language in order to exclude the uninitiated. But this accusation overlooks the central point. As mentioned above, algebra and symbolic logic give us a way of proceeding mechanically from assertion to assertion. In algebra, we use the *rules of algebra*, alluded to by Cardano in the epigraph to Part III, and in symbolic logic we use the *rules of inference*.

Coldly applying abstract rules of inference might seem to replace a meaningful problem with a mindless algebraic formalism, a meaningless game. On the contrary, the power of algebra is that it examines the abstract essence of a problem, ignoring confusing unessential details. In the sixteenth century, Cardano recognized the power of algebra and rightly called it *Ars Magna*, the Great Art.

The next three chapters are a brief introduction to the calculus, the mathematics of the indiscrete.

Part IV

A Smoother Pebble

If I have made any valuable discoveries, it has been owing more to patient attention than to any other talent.

—Isaac Newton

10

On the Shoulders of Giants

If I have seen further it is by standing on the shoulders of giants.

—ISAAC NEWTON, Letter to Robert Hooke

ALCULUS is Latin for *pebble*. The original meaning is preserved in medical terminology: a kidney stone is called a *renal calculus*. In antiquity, *calculi* were used for voting—black or white indicating condemnation or acquittal. However, the ancient use of pebbles for reckoning is the source of our common usage: a *calculus* is a method of computation. Specifically, *the* calculus refers to a method of computation developed in the seventeenth century. Truly, "method of computation" understates the nature and importance of the calculus, which opened up a new view of the world and led to Newton's astonishing discoveries in physics. Today, it is hard to find any corner of modern science or technology that does not rest on the foundation of the calculus.

Isaac Newton (1642–1727) discovered the calculus during the years 1665–66, apocalyptic years for London encompassing first the Great Plague and then the Great Fire. While Cambridge University was closed on account of the plague, Newton—only 23 years old—returned home to Woolsthorpe, where he continued his studies independently. Newton did not publish his discovery of the calculus until much later. In 1675, the calculus was independently rediscovered by the German mathematician/philosopher Gottfried Wilhelm Leibniz (1646–1716), who published his results in 1684. A bitter dispute over priority ensued between Newton, Leibniz, and their supporters. The discovery of the calculus is generally credited equally to Newton and Leibniz. In this chapter, we will see that parts of the calculus were discovered even before Newton and Leibniz.

The two central problems of the calculus are *the problem of areas* and *the problem of tangents*. In calculus and its applications these problems are generalized beyond areas and tangents; therefore, we have the technical names *integration* and *differentiation*, respectively. Roughly speaking, integration is a method of evaluating a quantity exactly by approximating it with sums of an ever larger number of ever smaller terms; and differentiation is the computation of the rate of change of one variable with respect to another. Integration includes the problem, for example, of finding the length of a curve, and differentiation includes the problem of finding the velocity of a particle. Although problems of integration and differentiation were considered before Newton and Leibniz, their genius consisted in greatly expanding the scope of these two ideas and, most of all, in finding the connection between them. In this chapter, we see how these problems were treated before Newton and Leibniz—integration by Archimedes (287?–212 BCE) and differentiation by Pierre de Fermat.

Integration Before Newton and Leibniz

Archimedes' method for estimating pi

Although popularly everyone called a Circle is deemed a Circle, yet among the better educated Classes it is known that no Circle is really a Circle, but only a Polygon with a very large number of very small sides.

—EDWIN A. ABBOTT, Flatland (1884)

Ancient Greek mathematicians made discoveries that foreshadowed the calculus, despite their lack of adequate mathematical notation. In particular, we will see how Archimedes found estimates of the number π. Archimedes used a technique of Eudoxus called *the method of exhaustion*; he approximated the circumference of a circle to any desired degree by inscribing polygons with a sufficiently large number of sides.

The ancient Greek geometers encountered a discontinuity at their first discussion of the circumference and area of a circle—as do high school geometry students of today. This is because the concepts of circumference and area of a circle are difficult to relate to lengths and areas of figures built from straight line segments—the only kind of figures previously considered. The length of a straight line segment is merely the distance between the two endpoints of the segment. The perimeter of a polygon is simply the sum of the lengths of its sides, but we need some additional concepts to apply this idea to the circumference of a circle.

Archimedes found an elegant method for computing the arc length of a circle and other curves. Using his method, Archimedes was able to prove that π, the ratio of the circumference to the diameter of a circle, is between $3\,10/71$ and $3\,1/7$. For simplicity, we will use his method to find a rougher approximation. However, his method can find approximations of π with any desired accuracy.

Pi, the ratio of the circumference to the diameter of a circle, evokes popular fascination. Perhaps π recalls our school days, or perhaps we try to see a hidden meaning in this infinite, nonrepeating decimal number. Pi has been the subject of errors and excesses. For example, in 1897 the Indiana House of Representatives voted unanimously to legislate an *incorrect* value of π.[1] Fortunately, the other house, the Indiana senate, tabled the matter and failed to enact it into law.

The British mathematician William Shanks (1812–82) spent many years computing by hand the value of π to 707 decimal places. In 1944, it was discovered that he had made an error at place 528, and everything from there on was incorrect. Today, a modest desktop computer can compute *correctly* the value of π to 707 decimal places in less than 10 seconds. The effort continues; π has now been computed to *billions* of decimal places. Other than demonstrating the raw power of modern computers, is anything useful accomplished by such computations? Computing the decimal digits of π investigates the question, Is π a *normal* number? A number is said to be *normal* if the distribution of its digits meets certain statistical tests of randomness.[2] The evidence to date supports the claim that π is a normal number, but no examination of even billions of decimal digits can provide a *proof*.

In school, I was taught to use the approximation $22/7$ for π; my grandfather used $355/113$—an approximation with an error of only 0.005 inches per mile. Clearly, the approximation that my grandfather used is adequate even now for the most exacting practical uses.

$$\frac{22}{7} = 3.1428\ldots \qquad\qquad 0.04\%\ \text{error}$$

$$\pi = 3.14159265\ldots \qquad\qquad \text{true value}$$

$$\frac{355}{113} = 3.14159292\ldots \qquad\qquad 0.000008\%\ \text{error}$$

Instead of *billions* of decimal places, we will compute a much rougher approximation for π.

Question 10.1. Show that the circumference C of a circle of diameter D is between $3D$ and $3.5D$. In other words, show that π is between 3 and 3.5.

Before we attempt to answer this question, we should back up a bit. Since our definition of π mentions the circumference of a circle, one might ask the following:

Question 10.2. What is meant by the circumference of a circle?

This question might seem an affectation of extreme skepticism, but let us humor the questioner. The way to give meaning to the word *circumference* is to describe in practical terms how we might measure the circumference of a circular object. The obvious answer is to use a flexible tape measure: wind it around the object, and take a reading. It's a sensible idea, but how do we proceed? Standard geometry doesn't have a useful abstract counterpart of the flexible measuring tape, but Archimedes had a bright idea that suffices.

Circular reasoning

Before we can discuss the circumference of a circle, we need to clarify the concept of arc length in general. In practical terms, we measure the arc length of a curve — in particular, the circumference of a circle — with a tape measure. At this time, we will not *define* the arc length of a circle or other curves. When we ask for the arc length of a circle of diameter 1, we are asking for a number that gives good agreement with the measured circumference using tape measures of arbitrary precision. We will arrive at such a number by *thinking* about the problem rather than making measurements with an *actual* tape measure. We begin by discussing arc length of curves of a special sort — convex curves.

The interior of a circle is called a *circular disk*. A circular disk is a special case of a *convex set*. A set is called convex if, for every two points P and Q in the set, the line segment PQ lies entirely within the set. Figures 10.1(a) and (b) are convex but Figures 10.1(c) and (d) are not convex. A curve that bounds a convex set, for example, a circle, is called a *convex curve*. The boundary curves in Figures 10.1(a) and (b) are convex curves. If we cut a convex figure (e.g., a circle or Figure 10.1(a) or (b)) from a piece of plywood and stretch a rubber band around the edge, the rubber band will assume the shape of the convex boundary curve without any gaps. The same is true if we cinch a tape measure tightly around the curve.

In Figure 10.2, the convex curve W is entirely inside the curve B. Think of W as the convex cross section of a waist and B as a loose-fitting belt. It is clear that the arc length of B is greater than the arc length of W because

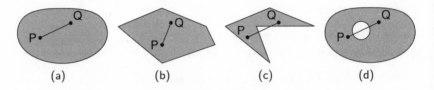

(a)   (b)   (c)   (d)

Figure 10.1. (a) and (b) are convex; (c) and (d) are not convex.

the slack in the belt can be cinched to fit the waist, thereby reducing the perimeter of the belt. We might even say, "It's a cinch" that the loose belt $\mathcal{B}$ will cover the convex waist $\mathcal{W}$.[3] We express this observation in the following axiom from Archimedes:[4]

Axiom 10.1. *If one convex curve contains another, then, unless the curves are identical, the inner curve has* shorter arc length *than the outer.*

Note that Axiom 10.1 is more subtle than its counterpart—not needed in this discussion—in which we replace "shorter arc length" with "smaller area":

Axiom 10.2. *If one convex curve contains another, then, unless the curves are identical, the inner curve has* smaller area *than the outer.*

Figure 10.2.

Completing the estimate of pi

We answer Question 10.1 with the help of Figure 10.3(a), a circle together with inscribed and circumscribed regular hexagons, and the blowup in Figure 10.3(b). By Axiom 10.1, the arc length of the circumference of the circle is between the perimeters of the inscribed and circumscribed hexagons. The inscribed and circumscribed hexagons have perimeters $12\overline{CD}$ and $12\overline{AB}$, respectively. If we compute these quantities, we will find lower and upper bounds for the circumference of the circle.

The segment OA is the radius of the circle; we denote it r. Note that the angle BOA is 30°. We can proceed by using the fact that a 30°–60° right triangle has sides, for example, OA, AB, and OB, in the ratio $2 : 1 : \sqrt{3}$.[5] From this ratio and $\overline{OB} = \overline{OD} = r$, we find

$$\overline{CD} = 0.5r \qquad \overline{AB} = \frac{r}{\sqrt{3}} \approx 0.5773r$$

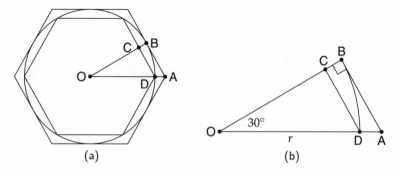

(a)   (b)

Figure 10.3.

Thus, noting $D = 2r$, the circumference of the circle in Figure 10.3(a) is between

$$12\overline{CD} = 6r = 3D \quad \text{and} \quad 12\overline{AB} \approx 6.9276r < 3.5D$$

We are finished; we have answered Question 10.1.

Differentiation Before Newton and Leibniz

A course in *coordinate geometry* (see page 93) is often a prerequisite for *differential calculus*. Since Descartes and Fermat discovered coordinate geometry, they had the necessary background to also discover differential calculus. In fact, they both did so.

Descartes and Fermat found different ways to use the methods of coordinate geometry to determine the direction of an arbitrary curve at an arbitrary point. In this sense, they were early pioneers in the study of differential calculus. They wrote about their discoveries in a manner that is hard for us to follow. This is true for two reasons: (1) They wrote without the benefit of today's standard terminology. (2) They were more concerned with breaking new ground than with presenting the simplest possible example for the reader. However, we are concerned with their *methods* rather than the particular form of their presentation. In the next two sections, we apply these methods to a certain problem. Although this specific problem was not considered by Descartes or Fermat, I believe that it shows quite simply the *discriminant* method of Descartes and the *difference quotient* method of Fermat. Discriminant and difference quotient are modern terms not used by Descartes or Fermat. We will use modern mathematical ideas as long as they simplify and do not obscure the central ideas of Descartes and Fermat.

The discriminant method was a clever initial solution to a difficult problem but was soon superseded by the simpler and more general difference quotient method. Broadly put, the problem is the following:

Problem 10.1. In Figure 10.4(a), find the direction of the curve C at an arbitrary point P.

Before attacking this problem, we must clarify several matters:

1. This is a problem of coordinate geometry. In order to proceed, we introduce the coordinate system shown in Figure 10.4(b). Descartes and Fermat used pairs of numbers to represent points in the plane, but the rectangular coordinate system with calibrated axes, as in Figure 10.4(b), is a modern concept that did not occur to them.

2. With respect to this coordinate system, we define the curve C by the equation $y = x - 0.25x^2$. It can be shown that C is a parabola, a conic section. Descartes and Fermat might have preferred to start by defining

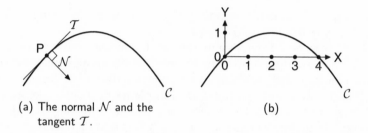

(a) The normal $\mathcal{N}$ and the
tangent $\mathcal{T}$.

(b)

Figure 10.4. The curve C is a parabola. In the coordinate system shown in (b), C has the equation $y = x - 0.25x^2$.

C *synthetically* as shown in Figure 5.21(b) on page 104,[6] but this would complicate our discussion without adding anything essential.

3. The direction of the curve C can be specified by means of either the *normal* $\mathcal{N}$, that is, the perpendicular, or the *tangent* $\mathcal{T}$ at the point P. Descartes used the normal and Fermat used the tangent.

Descartes's *discriminant* method

> *And I dare say that this is not only the most useful and most general problem in geometry that I know, but even that I have ever desired to know.*
>
> —RENÉ DESCARTES, La Géométrie

In the above quotation, Descartes refers to his discovery of a method of obtaining a *normal*, that is, a perpendicular, to a curve C at a specified point P. We will not discuss the details of Descartes's construction, but his general idea is to define a certain family of circles intersecting the curve C at the specified point P and at a second point Q. As the point Q approaches the point P, the corresponding circle approaches a tangent circle. A line from the center of this tangent circle to the point of tangency P is normal to the curve C at point P.

Descartes's use of a family of circles is a needless complication. It is simpler to use the family of *lines* that intersect P. As in Figure 10.5, a line $\mathcal{L}$ containing the point P generally intersects the curve C in a second point Q.

We use this simpler technique to solve Problem 10.1. In general, we do not feel constrained to use coordinate geometry precisely as Descartes understood it more than 360 years ago,

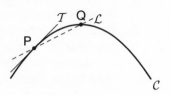

Figure 10.5.

but we retain Descartes's central idea: the use of the *discriminant*. We digress briefly to discuss this concept.

The *discriminant* provides a method to determine whether an algebraic equation has multiple roots. The discriminant is an algebraic expression that can be computed without solving the equation. The discriminant is zero if and only if the equation has multiple roots. Higher degree equations have discriminants, but here we will only consider the quadratic equation.[7] A quadratic equation generally has two distinct roots; for example, $x^2 - 3x + 2 = 0$ has roots $x = 1$ and $x = 2$. But in exceptional cases there may be only one root: for example, $x^2 - 2x + 1$ has the root $x = 1$ and no other. This one root is called a double root.

The general quadratic equation

$$ax^2 + bx + c = 0 \tag{10.1}$$

is solved by the quadratic formula

$$x = \frac{-b \pm \sqrt{b^2 - 4ac}}{2a}$$

In the quadratic formula, the plus-or-minus sign ($\pm$) indicates that, in general, there are two solutions. However, if $b^2 - 4ac = 0$ the two solutions of equation (10.1) coalesce into one: $x = -b/(2a)$. The quantity $D = b^2 - 4ac$ is called the *discriminant* of the quadratic equation (10.1).

Referring to Figure 10.5, we will use the discriminant method to find a tangent line at a certain point P. In particular, we will find the slope of the tangent to C at the origin relative to the coordinate system in Figure 10.4(b); that is, we assume that P is the point with coordinates $(0, 0)$. The discriminant method then proceeds as follows:

1. Find an equation to determine the point P of intersection with the curve C of an arbitrary line $\mathcal{L}$ through point P. This equation happens to be quadratic. One of its roots determines point P, and the second root determines point Q. Here are the details of this construction:

   (a) Find the equation of an *arbitrary* line through point P: $y = mx$ where m is the slope of the line.[8] Only the vertical line through P fails to represented by a suitable choice of the slope m.

   (b) Find the points of intersection of this line ($y = mx$) and the parabola ($y = x - 0.25x^2$). We can find the x-coordinates of the intersection points by solving for x in the equation that is obtained by eliminating y between the equation of the parabola C and the equation of the line $\mathcal{L}$, that is, between the equations $y = x - 0.25x^2$ and $y = mx$. The result is the equation $mx = x - 0.25x^2$, which is equivalent to

$$0.25x^2 + (m - 1)x = 0 \tag{10.2}$$

2. Find the line $\mathcal{L}$ such that the discriminant of this equation is equal to 0. In that case, points P and Q coincide, and the line $\mathcal{L}$ coincides with the tangent $\mathcal{T}$ to the curve $\mathcal{C}$ at point P. Here are the details:

   (a) First we must calculate the discriminant. This is done by observing that equation (10.2) is the same as the general quadratic (10.1) with $a = 0.25$, $b = m - 1$, and $c = 0$. Thus, the discriminant is equal to

$$D = b^2 - 4ac = (m - 1)^2$$

   (b) Setting $D = 0$, we see that a double root occurs if $m = 1$. Alternatively, we see that equation (10.2) has roots

$$x = 0 \text{ and } x = 4(1 - m)$$

   It is easy to see that these roots are equal if $m = 1$.

   (c) As the point Q approaches the origin P, the secant[9] line $\mathcal{L}$ approaches the tangent at point P. Moreover, the slope m of the secant line $\mathcal{L}$ ($y = mx$), approaches the slope of the tangent line $\mathcal{T}$. This happens when $m = 1$ because then both roots are equal to 0. Thus tangency at $(0,0)$ occurs when the slope m is equal to 1.

To apply Descartes's method for other curves, we need to test whether other types of equations have multiple roots. There are discriminants for equations of higher degree, but they quickly become complicated. For example, the discriminant for the general cubic equation

$$ax^3 + bx^2 + cx + d = 0$$

is given by the formula

$$b^2c^2 - 4ac^3 - 4b^3d - 27a^2d^2 + 18abcd$$

Descartes found a clever method, and he was justifiably proud of himself for finding it. However, his method leads very quickly to extraordinary complications. Fermat, on the other hand, found a much better solution to this problem. In fact, Fermat's method is the standard method found today in every calculus textbook.

Fermat's *difference quotient* method

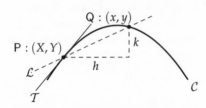

Figure 10.6. The slope of the secant line $\mathcal{L}$ tends to the slope of the tangent line $\mathcal{T}$ as h tends to 0.

We use modern terminology to describe Fermat's method; doing so does not obscure Fermat's central idea of *difference quotient*. We use Fermat's method to find the slope of the curve $\mathcal{C}$ in Figure 10.6 at the point P. We use (X, Y) to denote the coordinates of P with respect to the coordinate system shown in Figure 10.4(b). This problem is more general than one used to illustrate Descartes's method in the preceding section because here we find the tangent at an arbitrary point (X, Y) instead of the particular point $(0, 0)$.

To determine the slope of the tangent line through point P, Fermat's method begins, just as Descartes's did, by considering a secant $\mathcal{L}$ that intersects the curve $\mathcal{C}$ in the point P. The secant intersects the parabola a second time at point Q with coordinates that we denote (x, y). Here Fermat's method departs from Descartes's by naming the coordinates of this point. Let h be the difference in the x-coordinates of points P and Q, (i.e., $h = x - X$), and, similarly, for the difference in the y-coordinates put $k = y - Y$. The slope of the secant $\mathcal{L}$, the rise divided by the run, is equal to k/h. Notice that this fraction is meaningless if h is equal to 0. The key step of Fermat's method is to compute the difference quotient k/h ($h \neq 0$) as follows:

$$\frac{k}{h} = \frac{y - Y}{h} = \frac{\left(x - 0.25x^2\right) - \left(X - 0.25X^2\right)}{h}$$

Rearranging terms:

$$= \frac{(x - X) - 0.25\left(x^2 - X^2\right)}{h}$$

Eliminating x using $x = X + h$:

$$= \frac{h - 0.25\left[(X + h)^2 - X^2\right]}{h}$$

Using the identity $(X + h)^2 = X^2 + 2hX + h^2$:

$$= \frac{h - 0.25h(2X + h)}{h}$$

Canceling h from numerator and denominator:

$$= 1 - 0.5X - 0.25h \qquad (10.3)$$

As h tends to zero, formula (10.3), the slope of the secant $\mathcal{L}$, tends to the slope of the tangent $\mathcal{T}$ at the point P:

$$1 - 0.5X \qquad (10.4)$$

Note that this result is consistent with the result obtained using Descartes's method because, when X is equal to 0, formula (10.4) is equal to 1; in other words, the slope of the tangent to $\mathcal{C}$ at the point $(0,0)$ is equal to 1.

The preceding paragraph began, "As h *tends* to zero." Why is this circumlocution necessary? It seems simpler to say, "When h is *equal* to zero, formula (10.3) is *equal* to $1 - 0.5X$." The difficulty is that formula (10.3) is equal to the difference quotient of a secant line $\mathcal{L}$ *only if h is different from* 0 because, as remarked above, the difference quotient k/h is meaningful for arbitrarily small values of h but not for h equal to zero. If P and Q are distinct, then PQ determines a line, but a unique line is not determined if P and Q are identical.

In the seventeenth and eighteenth centuries, it was sometimes said that the slope of the tangent is precisely equal to a difference quotient k/h where h and k are "infinitesimally small." Newton and Leibniz and the other pioneers of calculus used infinitesimals quite freely, but they were called to task by Bishop Berkeley, who called infinitesimals "the ghosts of departed quantities." Berkeley's criticism did not get an adequate rebuttal until almost two centuries later—in the late nineteenth century—with the work of Dedekind, Kronecker, Weierstrass, and others.

Descartes and Fermat considered these findings an extension of geometry. Indeed, they did make exciting discoveries in geometry, but they failed to realize that their discoveries had importance far beyond geometry. Fermat was aware of Galileo's investigations concerning falling bodies and even attempted to rebut them. He failed to realize, not only that Galileo's assertions were correct, but even that they could be confirmed by using Fermat's own method, described above, for finding tangents to curves.

Galileo's Lute

Falling bodies

The ancient Greek philosopher Zeno of Elea (495?–430? BCE) asserted that the concept of motion leads to paradoxes. Aristotle also speculated on the nature of motion; however, the first scientific investigation of motion was undertaken by Galileo Galilei (1564–1642). As a young mathematics teacher at the University of Pisa during 1589–92, he is said to have performed a demonstration in which he dropped simultaneously two objects of different weight from a height of 117 feet atop the Leaning Tower of Pisa in order to show that the two objects fall at the same rate—contrary to the accepted dogma from Aristotle that the heavier one must fall faster. The authenticity of this story is controversial, but it is certain that some years later Galileo researched this question by rolling a ball down an inclined plane.[10]

The inclined plane has the advantage that it slows the descent, making it easier to measure subintervals of distance and time. Using the technology of the early seventeenth century, it was possible to measure distances but more difficult to measure small intervals of time. Sometimes Galileo used the volume of water flowing out of a tank to measure time, but, as we will see, he also used his musical sense of rhythm for this purpose.

Understandably, Galileo found instantaneous velocity a difficult concept because he lacked the knowledge of calculus needed to deal with this matter rigorously. Galileo initially rejected continuously varying velocity but later made it an important part of his theory of motion. It is not difficult to define instantaneous velocity for a motion of constant velocity. For one-dimensional motion, constant velocity is the distance traveled divided by time elapsed.[11] However, it is surprisingly difficult to define instantaneous velocity in general. The following paradox shows this difficulty.

Paradox 10.1. The velocity of a particle has no meaning for the following reasons:

- The velocity of the particle can only be determined by its position at various instants of time.
- The position of the particle at a given instant is insufficient to determine its velocity at that instant.
- The position of the particle at any other time has no bearing on the velocity of the particle at the given instant.

Velocity is a more subtle concept than one might expect. We rely on a speedometer to tell us that at a particular moment an accelerating automobile is traveling at 50 miles per hour, but what does this mean? The true explanation requires a consideration of the motion in arbitrarily small time intervals. The truth is that velocity requires the *differential calculus* for its very definition. Galileo had no knowledge of calculus. Fermat, 35 years younger than Galileo, discovered enough of the calculus to set this matter right, but he was unable or unwilling to make the leap from pure geometry to a question of motion. Instead of providing this key link, Fermat, who generally admired Galileo's work, mistakenly claimed that Galileo's results on falling bodies contained a contradiction.

For falling bodies, Galileo made certain assertions concerning the dependence on time of the distance fallen and the velocity. In fact, Galileo unknowingly solved a calculus problem. Let us now examine in more detail one of Galileo's beautiful experiments with an inclined plane.

The inclined plane

Galileo's father was a noted musician, a lutenist. In the following experiment, Galileo borrowed two concepts from the musical world: the frets of the lute and the rhythmic beat. Galileo's inclined plane, his "lute," is shown in Figure 10.7. He "tuned" the lute by adjusting the placement of catgut "frets" at points A–F along this inclined plane so that a ball released at O would make a noise as it rolled over each fret. Galileo adjusted the positions of the frets A–F until he heard a regular beat—in other words, until the time of passage from the start O to the first fret A was same as the time of passage between any other pair of adjacent frets. Perhaps he danced a jig to keep time with the ball.

The result of this experiment was astonishing, undoubtedly a high point of Galileo's scientific life. The result had an elegant regularity beyond anyone's expectation. A regular beat was heard when, apart from experimental error, the distances between adjacent frets was the sequence of odd multiples of the distance OA between the first two frets: 1, 3, 5, . . .

For this experiment it is convenient to use special units of time and distance. The unit of distance is the distance OA between the starting point and the first fret, and the unit of time is the time for the ball to roll between any adjacent pair of frets.

Galileo observed that the distances of the frets A–F from the starting point O were proportional to the squares of the integers:

$$\overline{OA} = 1 = 1^2$$
$$\overline{OB} = 1 + 3 = 4 = 2^2$$
$$\overline{OC} = 1 + 3 + 5 = 9 = 3^2$$
$$\overline{OD} = 1 + 3 + 5 + 7 = 16 = 4^2$$
$$\overline{OE} = 1 + 3 + 5 + 7 + 9 = 25 = 5^2$$
$$\overline{OF} = 1 + 3 + 5 + 7 + 9 + 11 = 36 = 6^2$$

As a result of this and other experiments, Galileo asserted a remarkable generalization: *When a body falls from rest, the distance descended is proportional to the square of the time of descent.* Using algebraic notation unknown to Galileo, if s is the distance descended and t is the time of descent, then $s = ct^2$ where c is a constant of proportionality.

A few years after this experiment, Galileo came to the correct conclusion that the *instantaneous velocity* of a falling body initially at rest was

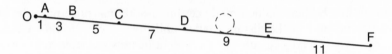

Figure 10.7. Galileo's lute: the inclined plane experiment. The ball starts from rest at the upper left point O. The ball hits the frets A–F at equal time intervals if the spacing between the frets is proportional to the odd integers: 1, 3, 5, 7, 9, 11, ...

proportional to the time elapsed. It is important to observe that Galileo did not make additional experiments devoted specifically to determining the velocity of a falling body. Galileo's instinct was correct. As we will see, the differential calculus shows precisely how the dependence of distance on time implies the dependence of velocity on time.

Galileo followed a tortuous path to obtain this result concerning the velocity of a falling body. It is possible that Galileo overlooked certain clues in the experiment of Figure 10.7 that could have made it easier for him to reach his conclusion regarding the velocity. Consider the following line of reasoning:

1. Instantaneous velocity is difficult, but *mean velocity* is much easier. To find mean velocity, divide the distance traveled by the time elapsed. For example, referring to Figure 10.7, to find the mean velocity of the ball as it rolls from the start O to the second fret B, divide the distance traveled, $1 + 3 = 4$, by the elapsed time, 2 beats. Dividing, we obtain velocity equal to $4/2 = 2$. (This calculation is also shown in the first row of Table 10.)

2. The mean velocity above over the time interval between the starting moment and the moment the ball hits fret B could be a first guess for the instantaneous velocity at the midpoint of this time interval — the moment of the first beat. (Galileo had no way to know that this first guess happens to be the *precise* value of the instantaneous velocity. In fact, when motion is governed by a quadratic expression like $s = ct^2$, as it is in the present case, the mean velocity over a time interval can be shown to be exactly equal to the instantaneous velocity at the midpoint of the time interval.)

3. Compute mean velocity over each time interval two beats in length: OB, AC, BD, CE, and DF. Table 10 shows this computation. The fact that the right column of this table is an arithmetic progression (2, 4, 6, 8, 10) supports Galileo's assertion that the velocity of a falling body is proportional to the time elapsed.

How Fermat could have helped Galileo

First things first! Fermat would have to begin by imparting some of his knowledge of algebra and coordinate geometry to Galileo. I feel sure that Galileo would have seen very quickly the importance of this knowledge.

Table 10.1. Computation of mean velocities for the inclined plane experiment in Figure 10.7. Mean velocity is computed for passage over pairs of adjacent fret intervals. This computation shows that the mean velocities 2, 4, 6, 8, 10 increase in an arithmetic progression. These mean velocities are reasonable estimates for the instantaneous velocities at the middle frets shown in the table. (In fact, it can be shown that these mean velocities are exactly the instantaneous velocities as the ball crosses the frets A–E.) Since the frets A–F occur at equal time intervals, the above computation supports Galileo's assertion that the velocity of a falling body is proportional to the time elapsed since the rest position.

	Start		Middle		End		
Interval	Time	Dist.	Time	Fret	Time	Dist.	Mean velocity
O–B	0	0	1	A	2	4	$4-0/2 = 2$
A–C	1	1	2	B	3	9	$9-1/2 = 4$
B–D	2	4	3	C	4	16	$16-4/2 = 6$
C–E	3	9	4	D	5	25	$25-9/2 = 8$
D–F	4	16	5	E	6	36	$36-16/2 = 10$

If Galileo and Fermat had had some further discussions, science and mathematics might have enjoyed a significant boost. Together, they might have seen that Galileo's experiments with falling bodies had an important connection with Fermat's method for determining tangents to curves. They might have seen that the mean velocities used by Galileo are formally the same as Fermat's difference quotients. They might have realized that a curve can represent a relationship between distance and *time*. For example, they might have been able to represent Galileo's result concerning falling bodies by means of a *graph* like Figure 10.8(a).

In fact, Galileo, Fermat, and their contemporaries did not conceive of such a graph. How is that possible when today a graph like Figure 10.8(a) seems so ordinary—even trite? It may help us to appreciate that this concept was an important abstraction if we realize that, in the jargon of Einstein's theory of relativity, the graphs in Figure 10.8 represent the motion of the falling body as a *world line* in *space-time* coordinates.

The calculation of instantaneous velocity is essentially the same as Fermat's difference quotient method (page 176) for determining tangents. Suppose we wish to determine the velocity at time T, that is, at the point P in Figure 10.8(b). We find the *average* velocity over an arbitrary time interval from T to t. Referring to Figure 10.8(b), the mean velocity is the same as the difference quotient:

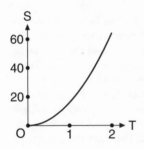

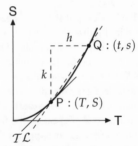

(a) Graph of s = 16t², the
relation between distance s in
feet and time t in seconds for
a falling body.

(b) Over the time interval (T, t)
of length h, the body falls
distance k. Mean velocity
equals k/h.

Figure 10.8. Graph of distance versus time for a falling body. The curve is a parabola.

$$
\begin{aligned}
\frac{k}{h} &= \frac{s - S}{h} = \frac{16t^2 - 16T^2}{h} \\
&= \frac{16(T + h)^2 - 16T^2}{h} \\
&= \frac{32hT + 16h^2}{h} \\
&= 32T + 16h
\end{aligned}
\tag{10.5}
$$

From (10.5), we see that as h, the length of the time interval, tends to 0, the mean velocity tends to $32T$, the instantaneous velocity at time T. This formula for instantaneous velocity confirms Galileo's assertion that the instantaneous velocity is proportional to the elapsed time T. Galileo reached this conclusion without any knowledge of calculus. In this sense, he solved a calculus problem unknowingly.

If the graph in Figure 10.8(b) were a straight line, then the mean velocity would be the same as the instantaneous velocity. This graph is a parabola, not a straight line, but the graph is smooth enough that in a 10-fold magnification of a neighborhood of point P, as shown in Figure 10.9, the curve appears straight. Under this magnification, the eye cannot distinguish between the curve and the tangent to the curve. This appearance of straightness under magnification is made precise by Fermat's difference quotient method.

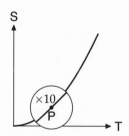

Figure 10.9. A neighborhood of point P on the graph in Figure 10.8(b) appears straight under a 10-fold magnification.

These early mathematical pioneers glimpsed the fundamental concepts of the calculus. They had difficulty in seeing the importance of their discoveries beyond the realm of geometry, and, more important, they failed to see the connection between the processes of integration and differentiation. We will explore this connection in the next chapter.

11

Six-Minute Calculus

God ever geometrizes.

—PLATO

God ever arithmetizes.

—CARL JACOBI (1804–51)

ARAPHRASING PLATO—nay, contradicting him—Jacobi could be expressing the new central role of the real number system in mathematics that came about at the beginning of the eighteenth century.
Euclidean geometry gave way to mathematical analysis—the calculus and
its higher ramifications—as the principal mathematical tool for understanding the world. Plato's view of a perfect universe gave way to the
modern view—a universe of endless detail and complexity that is understood only in various special contexts and only partially. The calculus was
born in the latter half of the seventeenth century and became a confident
young adult by the beginning of the eighteenth. Calculus could thrive
when mathematicians finally grasped the real number system. Their use
of the number system was pragmatic—the rigorous foundations came late
in the nineteenth century. Newton preferred geometry; nevertheless, his
work led to mathematics in which the arithmetic of the real numbers held,
and continues to hold, the dominant position. Newton showed us a universe that runs numerically instead of geometrically.

The real numbers work well for us because we live in a world in which
most physical processes seem to occur smoothly. Time and distance appear to be infinitely divisible. Calculus models the smooth physical processes that we see around us—from the steam rising from a teacup to
the motion of the planets. I wonder what sort of mathematics we would
have if quantum effects were of sufficient magnitude to be observed in
ordinary life.

In the eighteenth century, calculus was known as the *infinitesimal calculus* because it was—and still is—concerned with infinitely divisible quantities. Calculus draws conclusions by examining smaller and smaller quantities—for example, decreasing errors of approximation. Magically, sometimes calculus is able to leap from the approximate to the exact.

In the early history of calculus, infinitesimally small quantities were used uncritically. Using this incorrect notion—the only true infinitesimal is zero—Newton and Leibniz nevertheless found much correct science. The complete understanding of the foundations of calculus was not achieved until the late nineteenth century. In fact, during the 200 years after its discovery by Newton and Leibniz, calculus flourished despite unresolved issues concerning its foundations.[1]

The usual introductory course in calculus teaches a complex computational skill requiring a long apprenticeship. Although we cannot ignore this skill, here we are more concerned with the *meaning* of calculus. In scientific applications of calculus, a broad understanding of the meaning of calculus is often more important than computational skill. Acquiring computational skill in calculus is like learning a computer language. It is much more difficult to become a competent C programmer than to learn in general what computer languages are. Learning calculus skills is a substantial undertaking requiring at least a year of daily study and practice, but learning what calculus *is* can be accomplished much more easily. In fact, I hope to tell that story in the remaining pages of this book.

In the preceding chapter, we considered examples of integration by Archimedes and differentiation by Descartes and Fermat. Newton and Leibniz, who came later, are considered the discoverers of the calculus, largely because they saw that integration and differentiation are *inverse* operations. The details of this connection will be elaborated when we explore the *fundamental theorem of calculus.*

In this chapter, we consider differentiation and integration in the context of a six-minute automobile trip. First, we look at some preliminary matters: functions, limits, and continuity.

Preliminaries

In the last chapter, we saw that Descartes and Fermat foreshadowed the calculus in the context of analytic geometry by developing methods for finding normals and tangents to curves. They failed to see that these results had importance far beyond geometry. One important obstacle was the lack of a general concept of dependence of one variable on another—dependence of distance on time, volume on pressure, or any other dependence. Without question, Newton and Leibniz understood this concept. Newton called such a dependence a *fluent*, and today it is called a *function*.

Functions

A mathematical function expresses in a precise and general manner how the values of one variable determine the values of another variable. For example, the length of the side of a square determines the area of the square. If we use x and y to denote the length of the side and the area, respectively, then the variables x and y are connected by the formula $y = x^2$. In general, there may or may not be a specific meaning, distance, time, and so on, for the numerical variables x and y. We say that the variable y depends on x or that y *is a function of x.*

We wish to avoid awkward expressions like "Height at time $T + h$" and "Velocity at time T." Since we will continue this sort of discussion, these expressions will only get worse unless we do something about it. This clumsiness is improved by a new notation. We put $H(t)$ and $V(t)$ for the height and velocity at time t. Instead of "Height at time $T + h$," we write $H(T + h)$.

The mathematical meaning of the word *function* is similar to its ordinary meaning in the sentence, "The level of excellence is a *function* of the degree of preparation." (Practice makes perfect.) When the value of a variable x is determined by the value of another variable t, then we say that x is a *function* of t; x and t are called *dependent* and *independent* variables, respectively. A function of t can be denoted $L(t)$, $V(t)$, $f(t)$, and so forth. A function is a mathematical vending machine; for example, the squaring function (Figure 11.1) is a machine that accepts a number and gives in return the square of the number.

The input number (or expression) is called the *argument* of the function. The notation $f(\cdot)$ represents a function without reference to a particular argument. The set of possible numerical arguments, that is, the set of numbers x such that $f(x)$ is defined, is called the *domain* of the function f. The function defined by the equation $f(x) = x^2$ has a *natural* domain consisting of all real numbers. This could be made explicit with the definition "$f(x) = x^2$ for all real numbers x." We can define a function $g(\cdot)$ that is the same as $f(\cdot)$ but with a different domain, for example, the formula "$g(x) = x^2$ for real numbers x greater than 2," the squaring function with a domain consisting of numbers greater than 2 only. The functions $f(x)$ and $g(x)$ are considered different functions. The set of all image points of a function is called its *range*. For example, the range of the function f is

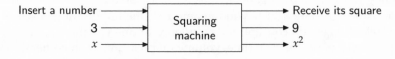

Figure 11.1. The squaring machine.

the set of nonnegative real numbers, and the range of g is the set of real numbers greater than 4.

A function is sometimes called a *mapping*. The function defined by the formula $y = x^2$ maps, for example, the number 2 into the number 2^2, that is, into the number 4. This function maps each number into its square. The formula $x \mapsto x^2$ (read "x maps to x squared") is a way of defining the same function without mentioning a second variable y. One can introduce a name for this function/mapping; we could call it f and write $f : x \mapsto x^2$. Alternatively, the formula $f(x) = x^2$ defines the same function. Using this definition, one can verify the following formulas:

$$f(2) = 2^2 = 4, \qquad f(x-2) = (x-2)^2 = x^2 - 2x + 4$$
$$f(f(x)) = f(x^2) = x^4, \qquad f(1+f(2)) = f(1+4) = 5^2 = 25$$

Suppose $f(\cdot)$ and $g(\cdot)$ are defined by the equations $f(t) = t^2 - 1$ and $g(t) = 1/t$ with a domain of definition consisting of all nonzero values of t. One can verify the following

$$f(2) = 2^2 - 1 = 3$$
$$f(t-1) = (t-1)^2 - 1 = t^2 - 2t + 1 - 1 = t^2 - 2t$$
$$f(f(t)) = (t^2 - 1)^2 - 1 = t^4 - 2t^2 + 1 - 1 = t^4 - 2t^2$$
$$f(g(t)) = (1/t)^2 - 1 = \frac{1}{t^2} - 1$$
$$g(f(t)) = \frac{1}{t^2 - 1}$$

Today's notation for functions, for example, $f(x)$, was originated by the Swiss mathematician, Leonhard Euler (1707–83).[2]

Graph of a function

We have seen in Chapter 5 that the concept of *graph* became ubiquitous in science and mathematics in the twentieth century, but that it was almost unknown until the beginning of the nineteenth century. Why was such an important concept so slow in coming? The answer is that a graph represents a *function* geometrically. Before there could be graphs, there had to be functions. The function concept was known at the beginning of the eighteenth century, and at the end of that century functions entered popular culture in the form of the graphs that are so common today. Figure 11.2 is a graph of the function $f(x) = x^2$.

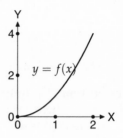

Figure 11.2. Graph of the function $f(x) = x^2$.

Limits

Limits express in a precise mathematical way the meaning of statements like

> *y tends to b as x tends to a.*

Here we assume that y is a function of x and that a and b are constants.

The above might seem an unnecessary circumlocution. Why not say instead,

> *y is equal to b when x is equal to a?*

There are two reasons why this simpler statement might not be correct:
1. The variable y might not be *defined* when x is equal to a. We will encounter this difficulty when we define the derivative of a function further below.
2. The variable y might tend to b as x tends to a, but the *value* of y at $x = a$ might be *different* from b.

To make these ideas concrete, we consider an example:

$$2x \text{ tends to } 4 \text{ as } x \text{ tends to } 2. \tag{11.1}$$

Is this the same as saying

$$2 \times 2 \text{ is equal to 4?} \tag{11.2}$$

In fact, these two assertions are *not* the same, but the distinction is subtle.

Putting the function $g(x)$ equal to $2x$ for all x, alternative forms for (11.1) and (11.2), respectively, are

$$\lim_{x \to 2} g(x) = 4 \tag{11.3}$$

and

$$g(2) = 4 \tag{11.4}$$

The definition of (11.1) and (11.3) is as follows:

Definition 11.1 (limit). Let the function $g(x)$ be defined in a neighborhood of $x = a$, where a is fixed. We say:
1. $g(x)$ tends to L as x tends to a; *or*
2. the limit of $g(x)$ as x tends to a is equal to L; *or*
3. $\lim_{x \to a} g(x) = L$;[3]

provided that $g(x)$ gets as close to L as desired whenever x is sufficiently close to a.

How is it possible to use Definition 11.1 to prove, for example, that formula (11.1) is correct? The proof consists in standing ready to rebut any challenge. For example, suppose someone asks,

> "How close to 2 does x have to be in order that the distance from $2x$ to 4 is less than 0.01?"

The answer is,

> "It is sufficient that the distance between 2 and x be less than 0.05,"

but this alone is not sufficient to establish (11.1). For every positive number e, however small, we must be able to answer the above question with 0.01 replaced by e. In our answer above, we only need to replace 0.05 with $e/2$.

The challenge and rebuttal implied by Definition 11.1 can be expressed in a more precise mathematical way. Two remarks prepare the way for this version of Definition 11.1. (1) There is a long-established tradition that, in this definition, one uses Greek letters for certain numerical variables: epsilon (ϵ) for the challenge and delta (δ) for the rebuttal. (2) This definition uses a rigorous method of stating that the values of two numerical variables are close together. For example, the formula $|p - q| < 0.01$ says that the gap between p and q is less than 0.01 without asserting which of the variables p or q is the larger.

A formal definition follows:

Definition 11.2. We say $\lim_{x \to a} g(x) = L$ if, for any $\epsilon > 0$, *however small*, there exists $\delta > 0$, *sufficiently small*, such that $|g(x) - L| < \epsilon$ (i.e., $g(x)$ is as close as desired to L) whenever $|x - a| < \delta$ (i.e., x is sufficiently close to a).

Continuity

Smoothness is an idea that underlies the calculus. *Continuity* is a weak form of smoothness. In the next section, we will see a stronger form of smoothness—*differentiability*. The graph of a continuous function can be drawn without lifting the pencil from the paper. However, we will see that the *definition* of continuity proceeds in a different direction.

In the preceding section, we defined $g(x) = 2x$ for all x. Let us define a closely related function $h(x)$. In fact, we define $h(x)$ to be the same as $g(x)$ except at $x = 2$. Here is a formal definition of $h(x)$:

$$h(x) = \begin{cases} 2x & \text{if } x \neq 2 \\ 0 & \text{if } x = 2 \end{cases}$$

The function $h(x)$, a trivial modification of $g(x)$, is not one of the leading citizens of the world of functions. It has no purpose other than to illustrate the concept of continuity—or perhaps I should say *discontinuity*, since $h(x)$ is discontinuous at $x = 2$. But I am getting ahead of the story. We must begin by defining what we mean by a continuous function.

Definition 11.3. Suppose that a function $f(x)$ is defined in a neighborhood of a. This function is said to be *continuous* at a if $\lim_{x \to a} f(x)$ exists and is equal to $f(a)$.

The function $h(x)$ fails to be continuous at $x = 2$. This is true because $\lim_{x \to a} h(x)$ is different from $h(0)$. Although this limit exists—in fact, it is equal to 4—its value is different from $h(0)$, which is equal to 0. The function $h(x)$ has the simplest kind of discontinuity at $x = 2$, a *removable* discontinuity, so named because redefining the function at a single point makes it continuous. Figure 11.3 shows graphs of two different kinds of nonremovable discontinuities at $x = 0$: the *jump* discontinuity and the *infinite* discontinuity.

As stated at the beginning of this section, the graph of continuous function can be drawn without lifting the pencil from the paper. However, this property is not contained in Definition 11.3. This fact is contained in the following theorem, together with information concerning the minimum and maximum values of a continuous function. This theorem is stated here and illustrated in Figure 11.4, but the proof is beyond the scope of this book.

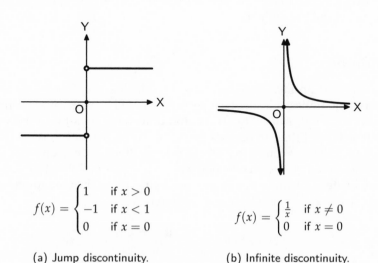

$$f(x) = \begin{cases} 1 & \text{if } x > 0 \\ -1 & \text{if } x < 1 \\ 0 & \text{if } x = 0 \end{cases}$$

$$f(x) = \begin{cases} \frac{1}{x} & \text{if } x \neq 0 \\ 0 & \text{if } x = 0 \end{cases}$$

(a) Jump discontinuity.   (b) Infinite discontinuity.

Figure 11.3. Two types of discontinuous functions.

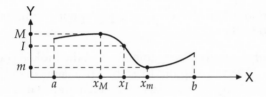

Figure 11.4.

Theorem 11.1. *Let $f(x)$ be continuous on an interval $a \leq x \leq b$.*

(a) **Minimum–maximum.** *Then there exists numbers x_m and x_M ($a \leq x_m, x_M \leq b$) such that $f(x_m) \leq f(x)$ and $f(x_M) \geq f(x)$ for all x between a and b, that is, for all x in the interval $a \leq x \leq b$.*

(b) **Intermediate value.** *Put $m = f(x_m)$ and $M = f(x_M)$, and let I (the intermediate value) be any number between m and M, that is, any number satisfying $m \leq I \leq M$. Then there exists at least one number x_I ($a \leq x_I \leq b$) such that $f(x_I) = I$.*

The next section introduces the two most fundamental concepts of the calculus, the derivative and the integral and discusses the connection between them, the *fundamental theorem of calculus*.

The Damaged Dashboard

Because it took great genius to create the calculus, one might think that it must deal with very abstruse matters, but this is not the case. We can illustrate the two central questions of calculus by discussing three devices found on every automobile dashboard: the *clock*, the *speedometer*, and the *odometer*.

A principal topic of calculus deals with the rate of change of one variable with respect to another—rates that Newton called *fluxions*. For example, the velocity of an automobile is the rate of change of distance with respect to time.

Question 11.1 (the damaged dashboard). If one of these three instruments breaks and the other two function accurately, can we estimate what the broken instrument should read?

No matter which instrument has failed, the answer is a qualified *yes*. The qualifications are as follows:

1. If one of the instruments is broken, we must make a careful log of readings on the remaining two instruments.
2. The accuracy of the result depends on
   (a) the accuracy and frequency of the readings

(b) the smoothness of the ride—expect poor results in erratic stop-and-go traffic

The cases of the broken speedometer and the broken odometer are of special interest. We will explore these two problems in detail. The problem of the broken clock is not discussed here because it is not used in the ensuing discussion of calculus. The problems of the broken speedometer and the broken odometer exemplify the *differential* calculus and the *integral* calculus, respectively.

The *damaged dashboard question* is easy if the velocity is constant because then the incremental distance (x) is equal to the rate (r) times the time (t). If two of the three variables x, r, and t are known, it is easy to compute the third from the relationship $x = rt$. Question 11.1 is of greater interest if the rate r is *not* constant. We emphasize this possibility by saying that the speedometer gives the *instantaneous* velocity, possibly different at each moment of time.

These two problems are illustrated by data involving a six-minute (0.1 hour) automobile journey. This trip is illustrated by the two graphs in Figure 11.5. These graphs entail the assumption that for each of the real numbers between 0 and 0.1, there is a corresponding moment of time, and there are real numbers giving the distance and velocity at this time. This an abstraction rather than a reality—a mathematical model that can be helpful in understanding an actual occurrence.

Distance and velocity as functions of time during the six-minute auto journey ($f(x)$ and $F(x)$, respectively, in Figure 11.5) are examples of continuous functions. An auto is unable to jump from point A to point B without passing through all of the intermediate locations. This commonsense observation accords with the continuity of the function $f(t)$ and Theorem 11.1(b).

Table 11 shows a log of readings of the speedometer, and the odometer taken at intervals of 0.01 hour (36 seconds). We will use parts of this

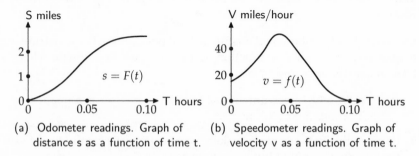

(a) Odometer readings. Graph of distance s as a function of time t.

(b) Speedometer readings. Graph of velocity v as a function of time t.

Figure 11.5. Graphs of distance and velocity for a six-minute journey. The horizontal axis measures time in hours: 0.01 hour = 36 seconds.

Table 11.1. Instrument log from the six-minute journey.

Clock Hours	Odometer Miles	Speedometer Miles/hour
0.00	0.0	15
0.01	0.2	21
0.02	0.4	30
0.03	0.8	42
0.04	1.3	51
0.05	1.8	45
0.06	2.1	33
0.07	2.4	21
0.08	2.6	9
0.09	2.6	3
0.10	2.6	0

log as a basis for approximate solutions of the broken speedometer and odometer problems.

The broken speedometer

The broken *speedometer* leads us to the *derivative*. Later, we will make a connection between the broken *odometer* and the *integral*. The derivative and the integral are the two central concepts of calculus.

The problem of the broken speedometer consists in using a log of data from the odometer to estimate the missing speedometer readings. This problem is illustrated in Table 11. In column 3, we find the *mean velocity*[4] in each time interval by dividing the distance traveled by the duration of the time interval. These mean velocities serve as estimates for the missing speedometer readings. For example, in column 3 the bold entry, 30, is the mean velocity in the time interval $(0.06, 0.07)$, obtained by dividing the distance $2.4 - 2.1 = 0.3$ by the duration 0.01. This mean velocity 30 is used as an estimate for the instantaneous velocity at time 0.06. (This is an arbitrary choice. The mean velocity could equally well be an estimate for the instantaneous velocity at 0.07 or any instant in the time interval $(0.06, 0.07)$.) The mean velocity 30 serves as an estimate for the missing speedometer reading, 33, in column 4.

Figure 11.6 shows the geometric meaning of the method in Table 11 for estimating the missing speedometer readings. In particular, Figure 11.6 illustrates that, in the bold row of Table 11, the estimate (30) for the missing speedometer reading is equivalent to using the slope of the line PQ as an estimate of the slope of the curve at point P (time = 0.06).

Table 11.2. The problem of the broken speedometer. The readings in the shaded column are unavailable. In column 3, the odometer readings are used to compute mean velocities, which estimate the missing speedometer readings of column 4.

1. Hours	2. Odometer	3. Mean Velocity	4. Speedometer
0.00	0.0	$(0.2 - 0.0)/0.01 = 20$	15
0.01	0.2	$(0.4 - 0.2)/0.01 = 20$	21
0.02	0.4	$(0.8 - 0.4)/0.01 = 40$	30
0.03	0.8	$(1.3 - 0.8)/0.01 = 50$	42
0.04	1.3	$(1.8 - 1.3)/0.01 = 50$	51
0.05	1.8	$(2.1 - 1.8)/0.01 = 40$	45
0.06	**2.1**	$(\mathbf{2.4 - 2.1})/\mathbf{0.01} = \mathbf{30}$	33
0.07	2.4	$(2.6 - 2.4)/0.01 = 20$	21
0.08	2.6	$(2.6 - 2.6)/0.01 = \ 0$	9
0.09	2.6	$(2.6 - 2.6)/0.02 = \ 0$	3
0.10	2.6		0

The accuracy of the estimates in the problem of the broken speedometer would be improved if odometer readings included more significant figures — for example, hundredths or thousandths of a mile instead of just tenths. Further improvements would result by using smaller time intervals. In fact, we achieve any desired degree of accuracy by using sufficiently small time intervals. Pursuing this idea we are led to the derivative.

The derivative

Calculus is the proper mathematical tool for understanding smooth and deterministic growth and motion. Rough and erratic motion, for example, the fluctuation of the stock market, is less natural and more difficult. Brow-

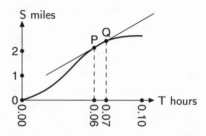

Figure 11.6. Geometric interpretation of the broken speedometer problem.

nian motion, the movement of tiny particles suspended in a fluid or gas subject to the constant bombardment of molecules, is a scientifically important example of erratic motion. On the other hand, the motion in a vacuum of a mass particle under the influence of gravity is an example of motion that is amenable to analysis using calculus.

The *derivative* generalizes the slope of the tangent to a curve and the instantaneous velocity of a moving particle. But keep in mind that some curves fail to have tangents everywhere, and a motion can be too erratic to have an instantaneous velocity. For example, instantaneous velocity is meaningless for Brownian motion. On the other hand, the motion of an automobile has an instantaneous velocity at every moment.

Figure 11.7 illustrates the concept of derivative of a function. The curve in this figure is the graph of $s = F(t)$, which expresses the dependence of the odometer readings on time during the previously discussed six-minute auto ride. The following makes no use of the specific nature of the function $F(t)$ except for the smoothness of the graph.

In Figure 11.7, the points P and Q represent the position of the auto at two instants of time, T and $T + \Delta t$, respectively. It is traditional to use the Greek capital delta, as in Δt or Δx, to represent a *change* or *increment* in the variable t or x.

Let us suppose that P in the figure is fixed and Q can move along the curve. In other words, the instant of time T is fixed, but duration Δt of the time interval can change. If Q is distinct from P, a secant line S is determined by PQ. As Q approaches P, that is, as the duration Δt of the time interval tends to 0, the secant S

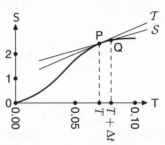

Figure 11.7. Graph of the function s = F(t) showing construction of the derivative at t = T.

approaches the tangent T. This is true because the graph of $F(t)$ is *smooth*; it would fail if the graph had sharp corners or jump discontinuities. Moreover, the ratio of rise to run, the slope of the secant line, approaches the slope of T, the tangent at P. We call the slope of the tangent T the *derivative* at point P.

Our use of the word *slope* needs clarification. The slope of the lines S and T is not slope in the usual geometric sense. The first complication is that run and rise are measured using the two different scales on the horizontal and vertical axes. Furthermore, the interpretation of s as distance and t as time means that here slope can be construed as *velocity*, distance per unit time. The slope of S is the mean velocity over a time interval of duration Δt, and the slope of T is the instantaneous velocity at time T.

For the increment in s occasioned by the increment Δt of t, put Δs. More specifically, Δs is equal to $F(T + \Delta t) - F(T)$. The mean velocity represented by the slope of $\mathcal{S}$ is

$$\frac{\Delta s}{\Delta t} \tag{11.5}$$

As Δt tends to 0, the mean velocity (11.5) tends to the instantaneous velocity at $t = T$, which we denote v_0. In other words, the difference between v_0 and the mean velocity (11.5) becomes as small as desired if the magnitude of Δt is sufficiently small. This relationship is also written

$$\lim_{h \to 0} \frac{\Delta s}{\Delta t} = v_0 \tag{11.6}$$

The fact that the difference quotient (11.5) tends to a limit as Δt tends to zero is a special property that might not hold for some functions. However, since $F(t)$ is the distance function for a physical motion, the limit of the difference quotient must exist. In fact, Newton's laws of motion imply that the distance function $F(t)$ has a continuous derivative. For some other function, the limit of the corresponding difference quotient might not exist as t tends to T. If the limit (11.6) exists, then we say that the *derivative of F exists* at T, that the function F is *differentiable* at T.

Here T represents an arbitrary instant of time between $t = 0.00$ and $t = 0.10$. However, the instantaneous velocity, the derivative $F'(t)$, exists for all t. In other words, like $F(t)$, $F'(t)$ is a function defined for all t between 0.00 and 0.10. The instantaneous velocity is also called the *derivative of the variable s with respect to the variable t*, denoted $F'(t)$ or ds/dt.

It is important to realize that, although ds/dt is the limit of the fraction $\Delta s/\Delta t$, it is not itself a fraction in the ordinary sense because we have not given an independent meaning to ds and dt. However, Leibniz, who first used this notation in 1675, believed (incorrectly) that ds and dt were infinitesimally small quantities that he called *differentials*.[5]

The derivative is meaningful apart from any physical meaning attached to the variables s and t. In fact, these variables and the derivative may be understood abstractly, without any physical meaning whatever. In this abstract setting, formula (11.5) is called a *difference quotient* instead of mean velocity or slope.

For an arbitrary function, the derivative may exist at some points and not at others. However, for the function $s = F(t)$, the derivative exists at every moment from $t = 0.00$ to $t = 0.10$. In other words, the function $F'(t)$, the velocity, is defined for t in the interval $(0.00, 0.10)$. Of course, the velocity is the function $v = f(t)$ shown in Figure 11.5(b). Therefore, we must have $f(t) = F'(t)$ for all t between 0 and 0.1.

The distance function $F(t)$ is an *antiderivative* of velocity function $f(t)$. The only other antiderivatives of $f(t)$ are of the form $F(t) + C$, where C is an arbitrary constant.

The readings of the speedometer and odometer are connected by the derivative–antiderivative relationship. The problem of the broken speedometer consists in using the odometer readings to approximate the derivative of the distance function.

Computation of derivatives

The relation between a function and its derivative generalizes the relation between the readings of the odometer and speedometer. A derivative is a *rate* of change of a dependent variable with respect to an independent variable. This definition does not explain why a course in differential calculus requires hundreds of hours of study. It appears that in this brief introduction we have omitted a great deal.

In particular, we have omitted any discussion of the computation of derivatives of functions that are defined by *formulas*. For example, the derivative of the function $f(x) = x^2$ is equal to $2x$. This can be seen as follows:

Recall that the derivative is the limit of the difference quotient, which for the function x^2 is equal to

$$\frac{(x + \Delta x)^2 - x^2}{\Delta x} \tag{11.7}$$

Algebraic simplification shows that formula (11.7) is equal to $2x + \Delta x$; we see that the limit of this expression as Δx tends to 0 is equal to $2x$. Thus, we have demonstrated the differentiation formula:

$$\frac{d}{dx} x^2 = 2x \tag{11.8}$$

Formula (11.8) is only the beginning. The standard course in differential calculus teaches powerful principles and techniques for finding derivatives of an amazing variety of functions. The following is a brief collection of rules for finding derivatives.

Recall that t^n has a meaning not only if n is an integer but even if n is an arbitrary real number—provided that t is positive. For example,

$$t^0 = 1 \qquad t^{-1} = 1/t \qquad t^{1/2} = \sqrt{t}$$

Thus, all of the items in Table 11.3 are special cases of the following formula in which n is an *arbitrary* real number.

$$\frac{dt^n}{dt} = nt^{n-1} \qquad (t > 0)$$

Table 11.4 gives rules for computing derivatives. Instead of proving these rules, we give two examples confirming the consistency of Figures 11.3 and 11.4.

Table 11.3. Table of derivatives. Item 8 is a consequence of the function of a function rule, below, together with items 3 and 7. Item 8 will be used on page 217.

		Function	Derivative	
1.	Constant	c	0	
2.	First power	t	1	
3.	Square	t^2	2t	
4.	Cube	t^3	$3t^2$	
5.	3/2 power	$t^{3/2}=t\sqrt{t}$	$\frac{3}{2}\sqrt{t}$	
6.	nth power	t^n	nt^{n-1}	$n \neq 0,\ t > 0$
7.	Reciprocal	$1/t$	$-1/t^2$	$t \neq 0$
8.	Square root	$\sqrt{t}$	$1/(2\sqrt{t})$	$t > 0$
9.		$\sqrt{a^2+x^2}$	$x/\sqrt{a^2+x^2}$	

Starting from the derivatives in Table 11.3, it is possible use the rules in Table 11.4 to compute derivatives of a large number of functions.

Example 11.1. From Table 11.3.3, we see that the derivative of t^2 with respect to t is equal to $2t$. Show that this result can be obtained from Table 11.3.2 and Table 11.4.3.

Solution. Use Table 11.4.3 with $f(t) = g(t) = t$. From Table 11.3.1, the derivative of t is equal to 1. Thus, we have

$$\frac{d}{dt}(t \cdot t) = t\frac{dt}{dt} + \frac{dt}{dt}t = 2t$$

Example 11.2. From Table 11.3.6, we see that the derivative of t^4 with respect to t is equal to $4t^3$. Show that this result can be obtained by using Table 11.3.3 and Table 11.4.4.

Table 11.4. Rules for derivatives. The functions $f(\cdot)$ and $g(\cdot)$ are arbitrary differentiable functions; c is a constant.

1.	Constant multiple	$\frac{d}{dt}cf(t) = cf'(t)$
2.	Sum of two functions	$\frac{d}{dt}(f(t) + g(t)) = f'(t) + g'(t)$
3.	Product of two functions	$\frac{d}{dt}(f(t)g(t)) = f'(t)g(t) + f(t)g'(t)$
4.	Function of a function	$\frac{d}{dt}f(g(t)) = \frac{d}{dt}f'(g(t))\,g'(t)$

Solution. Use Table 11.4.4 with $f(t) = g(t) = t^2$. Note that we have $t^4 = f(g(t))$. According to Table 11.4.4, we have

$$\frac{d}{dt}t^4 = \frac{d}{dt}(t^2)^2 = \frac{d}{dt}f(g(t)) = \frac{d}{dt}f'(g(t))g'(t)$$
$$= f'(t^2)(2t) = (2t^2)(2t) = 4t^3$$

Distance, velocity, and acceleration

As noted on page 179, Galileo discovered that the distance s traveled in time t by a falling body is proportional to the square of the time elapsed since rest. In other words, there exists a constant c such that $s = ct^2$. Velocity is the derivative of distance with respect to time, and acceleration is the derivative of velocity with respect to time. Thus, from Galileo's formula, we find that the velocity v is equal to $2ct$, and the acceleration a is a constant $2c$. Galileo might have thought it mysterious that distance is proportional to the square of the time, but here we see that the underlying reason is simple. A falling mass particle is subject to constant downward acceleration of gravitation.

According to Newton's *law of universal gravitation*, gravitational acceleration is *not* constant. Nevertheless, the conclusions of this section are correct *on the surface of the earth*, where Galileo made his experiments, and where the variation in gravitational acceleration is negligible. On the earth's surface, the constant gravitational acceleration is generally denoted g and is approximately equal to 32 feet (980 centimeters) per second per second.

Galileo made a further observation—that the *velocity* of a falling body is proportional to the square root of distance traveled from rest. This fact can be found by eliminating t from the two equations $s = \frac{1}{2}gt^2$ and $v = gt$. This basic fact also can be derived as a consequence of the law of *conservation of energy*. It can be shown—not only for linear motion, but also for arbitrary curvilinear motion—that v, the magnitude[6] of the velocity of a falling mass particle, is proportional to the square root of the vertical distance y below the rest position: $v = \sqrt{2gy}$. This equation describes the motion (ignoring the effect of friction) of a bead sliding on a curved wire or a roller coaster on an involuted track.

The broken odometer

The broken *odometer* leads to the *integral*. As remarked above, the derivative and the integral are the two central ideas of calculus.

The problem of the broken odometer consists in using a log of data from the speedometer to estimate the missing odometer readings. This problem is illustrated in Table 11. The first two columns are the log of

Table 11.5. The problem of the broken odometer. The readings in the shaded column (column 6) are unavailable. In column 3, the speedometer readings are used to compute average velocities, which are used to compute the estimated incremental distances in column 4. Total distances in column 5 estimate the missing odometer readings of column 6.

1. Hours	2. Speedo-meter	3. Average Velocity	4. Incr. Miles	5. Total Miles	6. Odo-meter
0.00	15			0.000	0.0
0.01	21	$(15+21)/2 = 18.0$	0.180	0.180	0.2
0.02	30	$(21+30)/2 = 25.5$	0.255	0.435	0.4
0.03	**42**	$(30+42)/2 = 36.0$	**0.360**	**0.795**	**0.8**
0.04	51	$(42+51)/2 = 46.5$	0.465	1.260	1.3
0.05	45	$(51+45)/2 = 48.0$	0.480	1.740	1.8
0.06	33	$(45+33)/2 = 39.0$	0.390	2.130	2.1
0.07	21	$(33+21)/2 = 27.0$	0.270	2.400	2.4
0.08	9	$(21+9)/2 = 15.0$	0.150	2.550	2.6
0.09	3	$(9+3)/2 = 6.0$	0.060	2.610	2.6
0.10	0	$(3+0)/2 = 1.5$	0.015	2.625	2.6

the clock and speedometer readings. The missing odometer readings are shown shaded in the last column. In Table 11, we estimate the missing odometer readings as follows:

Column 3: In each time interval, compute the arithmetic mean of the starting and ending velocities. We use the term *average velocity* for this value. Note that we have previously used *mean* velocity in a different sense. For example, see Table 10.

Column 4: Use these average velocities to estimate the incremental distance traveled during each time interval.

Column 5: Find the cumulative totals to estimate the missing odometer readings.

Notice that the estimated odometer readings in column 5 are in good agreement with the actual odometer readings in the last column. These estimates could be improved even more by including more frequent speedometer readings in the log.

Figure 11.8 illustrates the above procedure for estimating the missing odometer readings. This figure consists of Figure 11.5(b) with 10 vertical rectangles inserted. For example, the shaded rectangle corresponds to the bold row of Figure 11.8. The base of the shaded rectangle is the time interval $(0.02, 0.03)$, and the height of this rectangle is 36.0, the average velocity in the time interval. The *area* of a rectangle is its width multiplied by its height, $0.01 \times 36.0 = 0.36$. But this is not area in the ordinary sense. For one thing, the horizontal scale is different from the vertical scale. More important, we are multiplying horizontal units of time by vertical units of velocity, that is, distance per unit time. Therefore, the product must have units of distance, and, in fact, this product, 0.36 miles, is an estimate for the distance traveled in the time interval $(0.02, 0.03)$.

In Table 11, the estimate 0.795 (bold in column 5) for the missing odometer reading at time 0.03 is equal in Figure 11.8 to the sum of the areas of the first three rectangles, the gray rectangle and the two rectangles to its left.

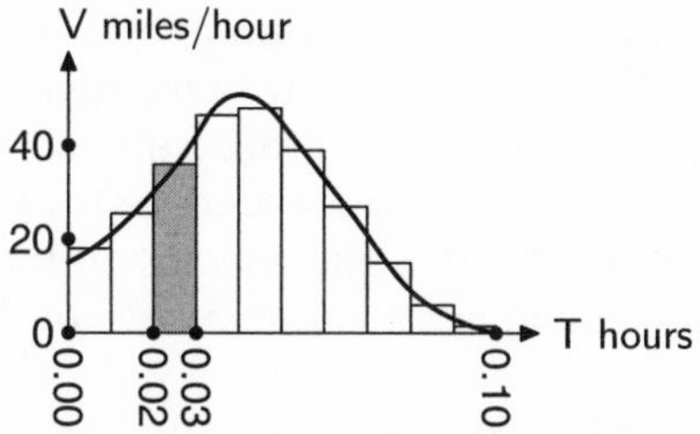

1. The use of the arithmetic mean to compute the average velocities in column 3 of Table 11 is an arbitrary choice. Other methods are possible for finding an intermediate velocity in each time interval.

Figure 11.8. Geometric interpretation of the broken odometer problem. The area of the shaded rectangle is equal to the bold item in column 4 of Table 11.

2. The estimates in column 5 can be interpreted as estimates for the area under the speedometer curve in Figure 11.5(b).

The definite integral

The definite integral is a concept that enables us to define in Figure 11.5(b) the area bounded by the curve $v = f(t)$ and the horizontal and vertical coordinate axes. In the preceding section, this area was approximated by rectangles, as shown in Figure 11.8. Refining this method of approximation tends, in the limit, to the precise value of this area. However, the definite integral is a concept of great generality that goes far beyond areas.

The weight of a string of beads is equal to the sum of the weights of the individual beads. A generalization of this problem is to find the weight of a nonhomogeneous wire with a nonconstant linear density (e.g., grams per centimeter). To find the total weight of the wire from a given density function, we must generalize the sum of the weights of the string of beads. Fancifully—for the moment leaving mathematical rigor behind—we might think that the wire is composed of infinitely many infinitesimally small beads. The total weight of the wire is an example of a mathematical concept known as the definite integral.

A similar example is the problem of finding the total distance traveled by a particle that moves with a velocity that is not constant. The problem is to find the total distance given the velocity as a function of time. Indeed, the problem of the broken odometer requires such a computation. The solution of this problem in the preceding section may be satisfactory for any practical need, but the integral is more than merely an approximation method. In this section estimates of increasing precision lead to a new concept — the definite integral, a cornerstone of the calculus.

The discussion in the preceding section of the problem of the broken odometer gives a method of approximating the distance function $F(t)$ given a velocity function $f(t)$. The definite integral gives a precise construction of the distance function $F(t)$ from the velocity function $f(t)$ because integration is the ultimate refinement of the approximation methods used for the problem of the broken odometer.

This section offers two for the price of one — two important mathematical ideas: first, the concept of the *definite integral*, and, second, the fact that the integral is the inverse of the derivative — the *fundamental theorem of calculus*.

We need to set aside our concern over practical difficulties relating to the precision of actual clocks, speedometers, and odometers. In practice we can only examine the speedometer a limited number of times during the six-minute automobile journey. Nevertheless, as a thought-experiment we consider the possibility of an unlimited number of observations in order to define the definite integral. A definition that requires dealing with an unlimited number of data points might seem too complicated to be of practical use. Fortunately, as we will see, the *fundamental theorem of calculus* provides a remarkable shortcut for computing definite integrals.

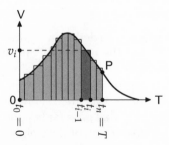

Figure 11.9. Graph of the function $v = f(t)$ showing construction of the definite integral from 0 to T.

To define the definite integral, the problem is this: *Given a continuous velocity function $v = f(t)$, for an arbitrary moment T $(0 \le T \le 0.1)$ find the distance $F(T)$ traveled between time 0 and time T.* The approximation method illustrated in Figure 11.8 is on the right track, but needs some refinement, illustrated in Figure 11.9.

Definition of the definite integral

We will formulate a procedure that gives the *exact* area of the shaded region bounded by the curve $v = f(t)$, the coordinate axes, and the vertical line through point P. This "area" is the distance $F(T)$ that the automobile travels in the time interval $0 \le t \le T$. (As noted in connection with Figure 11.8, there is ambiguity of our use of the word *area*. The ambiguity arises because of the contrast between the geometric *appearance* and the underlying *meaning* of Figures 11.8 and 11.9.)

This area is found as a limit of approximations, which are obtained by subdividing the interval $0 \leq t \leq T$. Figure 11.9 shows one such approximation in which the interval $0 \leq t \leq T$ is subdivided into n subintervals by means of the division points $t_0 = 0$, $t_1, \ldots, t_n$. Note that the subdivision points need not subdivide the interval $0 \leq t \leq T$ into subintervals of equal length. Although Figure 11.9 shows a subdivision into 12 subintervals, it is intended that n is an arbitrary positive integer. Furthermore, the ith subinterval is (t_{i-1}, t_i), where i is an arbitrary positive integer not greater than n.

Each of the n subintervals is the base of a tall thin rectangle. The sum of the areas of these rectangles is an approximation to the total area of the shaded region. The ith rectangle is shown in dark shading. (For each of the tall narrow approximating rectangles, we multiply the height, which has units of velocity, by the width, which has units of time. Velocity multiplied by time gives distance. Thus, the "area" in question is measured in units of distance. We will continue to use the word *area* in this way.)

For each i $(0 \leq i \leq n)$, the vertical height v_i of the ith approximating rectangle is the height of an arbitrarily chosen point on the curve above the ith subinterval (t_{i-1}, t_i). That is, the height of the rectangle with dark shading is $v_i = f(s_i)$, where s_i is an arbitrary number in the ith subinterval $(t_{i-1} \leq s_i \leq t_i)$.

Put $\Delta t_j = t_j - t_{j-1}$ $(0 < j \leq n)$, the width of the jth interval. With this notation, the sum S_n of the areas of all of the approximating rectangles is

$$S_n = f(s_1)\Delta t_1 + f(s_2)\Delta t_2 + \cdots + f(s_n)\Delta t_n$$

Using the *sigma notation*, the above can be written

$$S_n = \Sigma_{j=1}^{n} f(s_j)\Delta t_j \tag{11.9}$$

For a given subdivision of the interval $0 \leq t \leq T$, the smallest of the lengths of the subintervals is called the *mesh* of the subdivision.

If the sum S_n (11.9) of the areas tends to a value I as the mesh of the subdivision tends to 0 and n tends to ∞; that is, if S_n gets as close as desired to I provided that the mesh of the subdivision is sufficiently small, then we say that I is the *definite integral* of $f(t)$ for t between 0 and T, and, using a notation from Leibniz, we write

$$I = \int_0^T f(t)\, dt \tag{11.10}$$

It can be shown that if $f(t)$ is continuous, then the integral I exists. In other words, if $f(t)$ is continuous, the sum (11.9), regardless of the choice of the points s_i, tends to a limit as the mesh of the subdivision tends to 0. Later, we will have occasion to make a special choice of the points s_i.

This achieves our first goal, *the definition of the definite integral.* We continue now toward our second goal, to show *the sense in which differentiation and integration are inverse operations.* First, a few remarks concerning the definite integral:

1. The symbol $\int$ is from Leibniz. It is an elongated S, the first letter of the Latin word for sum, *summa.* Recalling that Σ is the Greek cognate of the Latin letter S, Leibniz's notation creates a similarity between formulas (11.9) and (11.10) that reminds us of the definition of the definite integral. The symbols $f(t)$ and dt in formula (11.10) are suggestive of the height and width of the tall narrow approximating rectangles.[7]

2. There are functions $f(t)$ so erratic that the definite integral (11.10) does not exist; that is, the sum S (11.9) does not tend to a limit as the mesh of the subdivision tends to 0. However, it can be shown that the definite integral always exists if $f(t)$ is a continuous function. If we assume that the velocity function $f(t)$ is continuous over the interval $0 \leq t \leq 0.1$ — a reasonable assumption for a velocity — then the definite integral (11.10) exists.

3. In the expression $\int_0^T f(t)\,dt$, 0 and T are called, respectively, the lower and upper limits of integration.

4. If the function $g(t)$ is defined for t in the interval $a \leq t \leq b$, then the definite integral $\int_a^b g(t)\,dt$ is defined similarly, and the integral exists if $g(t)$ is continuous.

The fundamental theorem of calculus

Until now we have not made use of the fact that $f(t)$ is the velocity function for the distance function $F(t)$ — in other words, $f(t)$ is the derivative $F'(t)$. Now we return to step 11 above.

Since s_i is chosen arbitrarily in the ith subinterval, we are free to make a special choice. In particular, in each subinterval we choose s_i so that $f(s_i)$ is the mean velocity for the ith time interval $(t_{i-1} \leq s_i \leq t_i)$.[8] The distance traveled in the ith time interval is equal to the mean velocity in that interval multiplied by the length of that time interval. In other words, we have

$$F(t_i) - F(t_{i-1}) = f(s_i)(t_i - t_{i-1}) = f(s_i)\Delta t_i$$

Now use this relation in the sum (11.9) to obtain

$$S_n = \Sigma_{j=1}^n f(s_j)\Delta t_j = \Sigma_{j=1}^n \left(F(t_j) - F(t_{j-1})\right)$$

The expression on the right is called a telescoping sum because all but two terms cancel. This can be seen more clearly if we write out the sum as follows:

$$(F(t_1) - F(t_0)) + (F(t_2) - F(t_1)) + \cdots + (F(t_n) - F(t_{n-1}))$$

The terms $F(t_1)$ cancel. In fact, all the terms of this sum cancel except

$$F(t_n) - F(t_0) = F(T) - F(0)$$

Thus, for every subdivision, the numbers s_i can be chosen so that the sum (11.9) is equal to $F(T) - F(0)$. Trivially, it is also true that the sum (11.9) tends to this same value as the mesh of the subdivision tends to 0. Thus, since $F(0) = 0$, we have shown

$$\int_0^T f(t)\,dt = F(T)$$

This form of the fundamental theorem of calculus is sometimes called the *first* fundamental theorem of calculus.[9]

It is usually stated more generally.

Theorem 11.2 (First fundamental theorem of calculus). *Let $g(t)$ and $G(t)$ be functions defined on the interval $I : a \leq t \leq b$. Moreover, suppose that $g(t)$ and $G(t)$ are continuous in I and that $G(t)$ is differentiable in the interior of I; and suppose $G'(t) = g(t)$. Then we have*

$$\int_a^b g(t)\,dt = G(b) - G(a)$$

On account of the relation $G'(t) = g(t)$, we say
- $g(t)$ is the derivative of $G(t)$, or
- $G(t)$ is an antiderivative (or indefinite integral) of $g(t)$.

We say "*an* antiderivative" because the antiderivative is not unique. The only other antiderivatives are obtained by adding an arbitrary constant, that is, $G(t) + C$ is an antiderivative of $g(t)$ for any constant C.

Thus, the problem of finding the definite integral of a function becomes the problem of finding antiderivatives. For example, from the formula

$$\frac{d}{dt}t^2 = 2t$$

that is, since t^2 is an antiderivative of $2t$, we obtain the definite integral

$$\int_2^3 2t\,dt = 3^2 - 2^2 = 5$$

A standard course in calculus devotes many weeks to the development of techniques for finding antiderivatives in special cases. It is much more difficult to find antiderivatives than to find derivatives. Functions like $\sqrt{1 + t^4}$ that can be defined using $+$, $-$, $\times$, $\div$, together with roots and powers, are called *elementary functions*. Derivatives of elementary functions are always elementary; however, there are many elementary functions whose antiderivatives are not elementary. For example, the antiderivative of the function $\sqrt{1 + t^4}$ is not elementary.

Roller Coasters

Roller coasters frighten and entertain by sending us down curved tracks of various shapes. The design of safe and entertaining roller coasters requires very sophisticated engineering. In this section, we apply calculus to study some highly simplified roller coasters.

The balls that fell down Galileo's inclined plane experienced such a ride, but the roller coaster is more exciting. The roller coaster pulls us slowly to rest at a great height. This is followed by a few seconds of free fall down a steep track. It is these few seconds of the ride that we will study.

Since roller-coaster tracks are generally curved, the first step is to determine the length of a curve—a curve in two dimensions because, for simplicity, we study only a track that is contained in a vertical plane.

The length of a curve

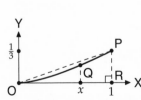

The preceding chapter discussed how Archimedes approached π by approximating the circumference of a circle. This was a step in the direction of the integral calculus. The rest of the story is that integral calculus can give us the length of an arbitrary smooth curve.

Figure 11.10. Graph of $y = \frac{1}{3}x\sqrt{x}$.

Problem 11.1. Find the length of the curve OP in Figure 11.10, the graph of $y = \frac{1}{3}x\sqrt{x}$ between $x = 0$ and $x = 1/3$.

Solution. We should find a result that is slightly longer than the length of the dashed straight line segment OP, which can be computed using the Pythagorean theorem. Applied to the right triangle OPR, this theorem asserts:

$$\overline{OP}^2 = \overline{OR}^2 + \overline{RP}^2$$

Using the numerical values for the lengths of these line segments, we have:

$$\overline{OP} = \sqrt{1^2 + \tfrac{1}{3}^2} = \sqrt{1 + \tfrac{1}{9}} \approx 1.05409$$

As shown in Figure 11.10, point Q corresponds to a certain value of the variable x. The arc length of the curve from O to Q is a certain function of x, which we denote $s(x)$.

We begin with a consideration of the derivative of $s(x)$, the limit of the difference quotient

$$\frac{s(x + \Delta x) - s(x)}{\Delta x}$$

For the numerator, $s(x + \Delta x) - s(x)$, put Δs. As usual, we put

$$\Delta y = y(x + \Delta x) - y(x)$$

The relation between Δx, Δy, and Δs is given, as shown in Figure 11.11 by the Pythagorean theorem:

$$\Delta s = \sqrt{\Delta x^2 + \Delta y^2}$$

Dividing by Δx, we obtain the difference quotient:

$$\frac{\Delta s}{\Delta x} = \sqrt{1 + \left(\frac{\Delta y}{\Delta x}\right)^2}$$

Figure 11.11. Δx, Δy, and Δs are governed by the Pythagorean theorem: $\Delta s^2 = \Delta x^2 + \Delta y^2$.

Finally, passing to the limit, we have (using the standard abbreviation y' in place of dy/dx).

$$\frac{ds}{dx} = \sqrt{1 + y'^2} \tag{11.11}$$

The value of $s(x)$ is the value of the integral

$$\int_0^x \frac{ds}{dx} dx = \int_0^x \sqrt{1 + y'^2} \, dx$$

This formula is quite general. We have not made use of any particulars of the curve. Now we use the fact that the curve is the graph of $y = \frac{1}{3}x\sqrt{x}$ between $x = 0$ and $x = 1$. First we use the table of derivatives, Table 11.3.5, to find dy/dx:

$$\frac{1}{3}\frac{d}{dx}x\sqrt{x} = \frac{1}{2}\sqrt{x}$$

Now we compute ds/dx:

$$\frac{ds}{dx} = \sqrt{1 + y'^2} = \sqrt{1 + (\tfrac{1}{2}\sqrt{x})^2} = \frac{1}{2}\sqrt{4 + x} \tag{11.12}$$

Finally, the total length of the curve, from O to P is equal to the integral

$$s(1) = \int_0^1 \frac{1}{2}\sqrt{4 + x} \, dx \tag{11.13}$$

To evaluate this integral, we must find an antiderivative of $\frac{1}{2}\sqrt{4+x}$. In fact, we have

$$\frac{1}{3}\frac{d}{dx}(4+x)^{3/2} = \frac{1}{2}\sqrt{4+x}$$

Thus, by the fundamental theorem of calculus, the integral (11.13) is equal to

$$\frac{1}{3}\left((4+1)\sqrt{4+1} - (4+0)\sqrt{4+0}\right) = \frac{1}{3}\left(5\sqrt{5} - 4\sqrt{4}\right) \approx 1.06011$$

Recalling that the length of the straight line segment OP is 1.05409, we see that the length of the curve OP exceeds this figure by about 0.6%. Looking at Figure 11.10, this relationship seems plausible, thereby, increasing our confidence in the above calculation.

The above example seems complicated enough, but it is the simplest nontrivial example that I could find to illustrate the computation of the length of a curve. The computation of arc length tends to involve awkward calculations. Tables of antiderivatives are useful, but numerical methods are available in any case. The arc length of an ellipse, a simple and familiar curve, leads to antiderivatives that cannot be solved without introducing a new class of functions—the elliptic functions. Nevertheless, the definite integral gives a way of discussing arc length and many other geometric and physical concepts.

Time of descent

Descent on a straight track

Galileo's experiment with the inclined plane—equivalent to the descent of a roller coaster on a straight track—leads to a problem that has a solution involving a definite integral.

In the following problem, and in several problems to follow dealing with descent of mass particles down ramps, I use a unit of distance equal to $2g$, where g is the acceleration of gravity at the earth's surface. This unit is about 64 feet, close to the length of a surveyor's *chain*, which is exactly 66 feet. I will call this unit (64 feet) a *short chain*.

Problem 11.2. As shown in Figure 10.7, find the time for a mass particle starting from rest to slide without friction down the inclined plane from O to P, a distance a downward and b forward. Show that the time of descent is the same as if the speed over the entire descent were constantly equal to one-half the final speed at point P.

Solution (first method). Figure 11.12 shows the usual X,Y coordinate system, except that downward y distances are considered positive. The diagonal length c of the inclined plane is given by the Pythagorean theorem:

$c = \sqrt{a^2 + b^2}$. When the particle reaches depth y its speed is $v = \sqrt{2gy}$, where g is the acceleration of gravity. (See page 199.) The speed v at depth y in *short chains* (64 foot units) is given by the simpler formula

$$v = \sqrt{y} \text{ short chains/sec} \qquad (11.14)$$

If s represents distance along the inclined plane OP, then the relation between y and s is given by $y = \frac{a}{c}s$ and the speed v is equal is given by

$$v = \sqrt{y} = \sqrt{\frac{a}{c}}\sqrt{s} \qquad (11.15)$$

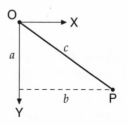

Figure 11.12. The inclined plane, $c = \sqrt{a^2 + b^2}$.

The time T of descent is the integral with respect to distance s of the reciprocal of the speed:

$$T = \int_0^c \frac{ds}{v} \qquad (11.16)$$

Use the expression (11.15), we find

$$T = \sqrt{\frac{c}{a}} \int_0^c \frac{1}{\sqrt{s}}\, ds \qquad (11.17)$$

From the table of derivatives, Table 11.3.8, and the fundamental theorem of calculus (Theorem 11.2), the definite integral (11.17) is equal to

$$\sqrt{\frac{c}{a}} \cdot 2\sqrt{c} = \frac{2c}{\sqrt{a}} \qquad (11.18)$$

This answer has a commonsense interpretation. The starting speed of the mass particle is 0, and the speed reaches $\sqrt{a}$ when the particle reaches the depth a at the end of the ramp. One might guess that the time of descent would be roughly the same if the speed had a constant value, throughout the entire descent, equal to the arithmetic mean of the starting and ending velocities: $\frac{1}{2}\sqrt{a}$.

In fact, the distance traveled, c, divided by the elapsed time, T, gives the *mean speed* over the time interval 0 to T, which we will call v_0. Using (11.18), we find

$$v_0 = \frac{c}{T} = \frac{c}{2c/\sqrt{a}} = \frac{1}{2}\sqrt{a}$$

But $\sqrt{a}$ is equal to the speed v_1 at point P, the bottom of the inclined plane. Thus, the *mean speed v_0 is exactly half of the maximum speed*, the speed at point P. This phenomenon was alluded to in the legend of Table 10 on page 181. This observation is the basis of a second solution:

Solution (second method). We start from the fact that for a particle — whether falling freely or sliding on an inclined plane — the relation between distance s and time t is given by $s = ct^2$, where c is a suitably chosen constant. Since the velocity is the derivative of distance with respect to time, the velocity v_1 at time T, at the end of the descent, is equal to $2cT$. On the other hand, the mean velocity v_0 is equal to

$$\frac{cT^2}{T} = cT = \tfrac{1}{2}v_1$$

Problem 11.3. In Figure 11.10, find the time of descent of a mass particle that slides down the dashed line segment from P to O.

Solution. The maximum speed, attained at O, is the square root of the height of the ramp: $\sqrt{1/3}$, and the mean speed is half that number. To find the time of descent T, we divide length of the ramp by the mean speed, obtaining

$$T = \frac{\sqrt{1^2 + \tfrac{1}{3}^2}}{\tfrac{1}{2}\sqrt{\tfrac{1}{3}}} \approx 3.65147 \text{ seconds}$$

Descent on a curved track

Galileo could have tried replacing the inclined plane with a curved ramp.

Figure 11.10 suggests a contest. In the preceding section, we showed that a mass particle, call it particle A, slides down the dashed straight ramp PO in 2.94391 seconds. Suppose that we slide particle B down the solid curved ramp PO. Which particle will win the race? The answer is not immediately clear. Particle B has an initial advantage because the solid ramp is steeper near point P. On the other hand, particle A has the advantage of a shorter path.

Problem 11.4. In Figure 11.10, find the time of descent of a mass particle that slides down the solid curve from P to O.

Solution. We measure length in *short chains* (64 foot units) as in Problem 11.2. The time T of descent is equal to

$$\int_0^1 \frac{ds}{v}$$

where s is arc length and v is the speed of the particle. It is more convenient to use the horizontal distance x as the variable of integration:

$$T = \int_0^1 \frac{1}{v}\frac{ds}{dx}\,dx \tag{11.19}$$

The speed v is the square root of the depth of the point Q. Since the descent starts from rest at point P $(x = 1)$, we have

$$v = \sqrt{\tfrac{1}{3}(1-x)}$$

Equation (11.12) established that ds/dx is equal to $\tfrac{1}{2}\sqrt{4+x}$. Substituting these expressions for v and ds/dx in equation (11.19), we find:

$$T = \int_0^1 \frac{\tfrac{1}{2}\sqrt{4+x}}{\sqrt{\tfrac{1}{3}(1-x)}}\,dx = \frac{\sqrt{3}}{2}\int_0^1 \sqrt{\frac{4+x}{1-x}}\,dx \qquad (11.20)$$

The evaluation of this integral could be a problem in a standard calculus course, or it could be achieved using a computer program such as *Mathematica* or *Maple*, but it is beyond the scope of this book. Readers who are calculus students may wish to verify that the integral (11.20) is approximately equal to 3.73970 seconds.

Recall that the time of descent for the straight track was 3.65147. The straight ramp wins the race by 0.08822 seconds—a margin of about 2%. Is the straight ramp unbeatable? In the next chapter we will see that this is not the case, that a curved track in the shape of a cycloid achieves the least possible time of descent.

12

Roller-Coaster Science

All is for the best in this best of all possible worlds.

—VOLTAIRE, Candide, 1759

THE OPTIMISM OF DOCTOR PANGLOSS is the butt of Voltaire's biting satire. Actually, Doctor Pangloss is a stand-in for Gottfried Leibniz, who enjoyed more esteem as co-discoverer of the calculus than for his philosophy of optimism. In physics, however, principles of optimality stand on firmer ground. In fact, there are several *variational principles* that state that of all conceivable happenings, nature chooses the one of least "effort." The oldest of these principles is known as *Fermat's principle of least time*, which states that a ray of light selects, from all possible paths between two points, the path of least time. Fermat's principle explains the reflection and refraction of light. In 1746, Pierre-Louis Moreau de Maupertuis (1698–1759) formulated a similar minimum principle for mechanics, the *principle of least action*, that characterizes the motion or equilibrium configuration of a mechanical system, for example, a pendulum, a gyroscope, or a suspension bridge. This principle says that, of all the conceivable configurations of a mechanical system, the one that actually occurs minimizes a certain quantity. A similar variational principle, *Hamilton's principle*, occurs in both classical and quantum mechanics. The search for all-inclusive variational principles is a chapter in the physicists' quest for a *theory of everything*.

The search for extrema—the best, worst, most, or least—is also of interest for more mundane problems. The calculus is often the best tool for solving such problems.

In this chapter, we examine how calculus is used to solve the simplest extremum problems. Then we will examine tools, inequalities, that can be used to solve extremum problems without calculus. Finally, we will

212

use inequalities to give a recent proof of the oldest variational problem, the problem of designing the fastest roller coaster, also called the *brachistochrone problem*.

The Simplest Extremum Problems

The simplest type of extremum problem asks for the maximum or minimum of a function. Such a problem can be reduced to finding the highest or lowest point on a graph. Theorem 11.1 established that a function $f(x)$, continuous on an interval $a \le x \le b$, has both a maximum and a minimum value in that interval. It often happens that $f(x)$ is a differentiable function, as in the following problems.

The rectangle of maximum area with fixed perimeter

Problem 12.1. Find a value of x between 0 and 1 such that $x - x^2$ is as large as possible.

Problem 12.2. Show that of all rectangles of fixed perimeter p, the square has the largest area.

Problem 12.1 is clearly a problem of the type under consideration, albeit a practice exercise without much intrinsic interest. Problem 12.2 seems more interesting because of its geometric content. In fact, it is a simple instance of the class of *isoperimetric problems*—problems that ask for a geometric figure of fixed perimeter that maximizes area subject to various additional conditions. According to ancient legend, Queen Dido of Carthage solved an isoperimetric problem to demarcate her kingdom using the skin of a bull, but that is another story. We will see that Problem 12.1 is equivalent to a special case of Problem 12.2.

To see this equivalence, formulate Problem 12.2 as follows: To avoid awkward fractions, define q, the quarter perimeter of the rectangle, by the formula $q = p/4$. If the length of a side of the rectangle is x, then the adjacent side has length $2q - x$, and the area is

$$A(x) = x(2q - x) = 2qx - x^2 \quad (12.1)$$

for all x between 0 and $2q$. Thus, problem 12.1 is equivalent to the special case of Problem 12.2 with $q = 1/2$, that is, $p = 2$.

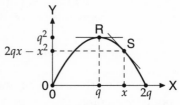

Figure 12.1. Problem 12.2: Find the rectangle of maximum area with fixed perimeter.

Figure 12.1 is a graph of the area y as a function of x; that is, $y = 2qx - x^2$. For x equal to 0 or $2q$ the rectangle degenerates to a line segment and the area is 0. We seek the point R where the area is maximum. We will prove shortly that the maximum value, q^2, is achieved at $x = q$.

The following theorem is the basis of a useful method for solving Problem 12.2 and similar extremum problems.

Theorem 12.1. *Let $f(x)$ be continuous in an interval $a \leq x \leq b$ and differentiable in its interior $a < x < b$. Suppose x_0 is a point such that $f(x_0)$ is either a maximum or minimum. Then either x_0 is a boundary point of the interval (i.e., $x_0 = a$ or $x_0 = b$) or the derivative $f'(x_0)$ is equal to 0.*

Proof. We consider the case of a maximum. (The following argument, with minor changes, also applies to a minimum.)

Theorem 11.1 asserts that a maximum of $f(x)$ must occur at some point x_0 in the interval $a \leq x \leq b$. Suppose that x_0 is an interior point of the interval, that is, that x_0 is different from a or b.

Recall that the derivative $f'(x_0)$ is equal to the limit as Δx tends to 0 of the difference quotient:[1]

$$\frac{f(x_0 + \Delta x) - f(x_0)}{\Delta x} \tag{12.2}$$

The crucial point is that the difference quotient approaches the same value from the right (Δx positive) as from the left (Δx negative). Since a maximum is achieved at $x = x_0$, it follows that $f(x_0 + \Delta x)$ cannot be greater than $f(x_0)$. In other words, the numerator of the difference quotient (12.2), $f(x_0 + \Delta x) - f(x_0)$, cannot be greater than 0. The difference quotient (12.2) is nonpositive (≤ 0) or nonnegative (≥ 0) depending on whether Δx is positive or negative. Therefore, $f'(x_0)$, the limit of the difference quotient (12.2), must be at the same time nonpositive and nonnegative. In other words, $f'(x_0)$ must be equal to 0. □

We continue with the solution of Problem 12.2—a model for the solution of many other similar extremum problems.

Solution (Problem 12.2). As shown above, Problem 12.2 is equivalent to finding the value of x in the interval $0 \leq x \leq 2q$ that maximizes the area function $A(x) = 2qx - x^2$.

According to Theorem 12.1, the maximum is achieved either at one of the boundary points ($x = 0$ or $x = 2q$) or at an interior point x_0 ($0 < x_0 < 2q$) such that $A'(x_0) = 0$. Since the area $A(x)$ is equal to 0 for $x = 0$ or $x = 2q$, neither of these values of x achieves the maximum. It follows that the maximum must be achieved at a point x_0 ($0 < x < 2q$) satisfying $A'(x_0) = 0$.

In general, there might be more than one point satisfying $A'(x_0) = 0$. In that case, we would need to check all such points and find the one that achieves the largest area. However, in the present case we will see that there is only one point x_0 $(0 < x_0 < 2q)$ such that $A'(x_0) = 0$.

Now we find the slope of the tangent at a point S corresponding to an arbitrary choice of x between 0 and $2q$ (see Figure 12.1). The slope is equal to the derivative:[2]

$$\frac{d}{dx}\left(2qx - x^2\right) = 2q - 2x$$

At the maximum point R in Figure 12.1, this slope must be 0. We find this point by solving the equation $2q - 2x = 0$, obtaining $x_0 = q$. For this value of x, the length of the second side of the rectangle is $2q - x_0 = 2q - q = q$. Thus, for maximum area, all four sides of the rectangle have length q; that is, the rectangle is a square.

The solution of Problem 12.1 is obtained by putting $2q = 1$. That is, the maximum occurs at $x = 1/2$.

Let us review two points in this proof that arise repeatedly in extremum problems:

1. *How do we know that the problem has a solution — that there exists a rectangle with maximum area?* The area (12.1) is a continuous function of x. The existence of a maximum area for x between 0 and $2q$ then follows from the intermediate value property of continuous functions — Theorem 11.1(b).

2. *How do we know that the value of x for which the derivative of (12.1) is 0 is actually a* maximum *for the area function?* According to Theorem 12.1, either the maximum of the area function $A(x)$ occurs for x equal to one of the endpoints of the interval $0 \le x \le 2q$, that is, at $x = 0$, or at $x = 2q$, or at an interior point of the interval, that is, a point x such that $0 < x < 2q$, where the derivative $A'(x)$ is 0. We find that there is just one such interior point, $x = q$. To find the maximum of $A(x)$, it is sufficient to examine three values: $A(0)$, $A(q)$, and $A(2q)$. Computing these three values, we find $A(0) = 0$, $A(q) = q^2$, and $A(2q) = 0$. The largest of these three, $A(q) = q^2$, is the required maximum. Therefore, the maximum area q^2 is achieved when $x = q$.

The lifeguard's calculation

In the next two sections we consider two problems that ask for a path of least time. The first of these problems is illustrated in Figure 12.2. This figure is a modification of Figure 11.12 because we will soon see a connection between Problem 11.2 discussed there and Problem 12.3 below.

Problem 12.3 (the lifeguard's calculation). A lifeguard is located at point P along the edge of the rectangular pool indicated by the shaded region in Figure 12.2. A drowning victim at O requires his attention. The lifeguard runs along the lower edge of the pool to point Q and then swims straight for the victim along the line segment QO. The lifeguard can run twice as

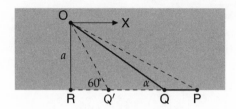

Figure 12.2. The lifeguard's calculation.

fast as he can swim. To reach the victim as soon as possible, how should he choose Q, his entry point into the pool?

Solution. We will see that he should use the following procedure: If the angle OPR is 60° or more, he should dive into the water where he stands and swim directly toward O. On the other hand, if the angle OPR is less than 60° he should run to the point Q′ such that the angle OQ′R is equal to 60° and then swim toward O.

The optimal angle 60° depends on the fact that the lifeguards runs exactly twice as fast as he swims. Another ratio for the speeds would entail a different optimal angle. However, the problem as stated relates nicely to the next problem.

Let x and b be the lengths of RQ and RP, respectively, and let v be the lifeguard's swimming speed. The time to run from P to Q is

$$\frac{b - x}{2v}$$

and the time to swim from Q to O is

$$\frac{\sqrt{x^2 + a^2}}{v}$$

Since v and b are constants, to minimize the sum of these two time intervals, it is sufficient to find the minimum of

$$y = 2\sqrt{x^2 + a^2} - x \qquad (12.3)$$

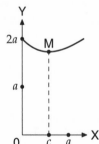

Figure 12.3. Graph of $y = 2\sqrt{x^2 + a^2} - x$. The minimum of y is achieved for x = c at the point M.

The graph of equation (12.3) is shown in Figure 12.3. The minimum of y is achieved where $x = c$ at point M. The graph indicates—and it can be shown rigorously—that y is a decreasing function of x when $x < c$ and increasing when $x > c$.

To find this minimum, that is, to compute the value of c, we first compute the derivative $y′$

$(= dy/dx)$ using item 9 in the table of derivatives, Table 11.3 on page 198:

$$y' = \frac{2x}{\sqrt{a^2 + x^2}} - 1$$

Setting y' equal to 0, we obtain the condition

$$\frac{x}{\sqrt{a^2 + x^2}} = \tfrac{1}{2}$$

The fraction on the left is the ratio two lengths in Figure 12.2, $\overline{RQ} : \overline{OQ}$. This ratio depends only on the angle α; in fact, it is equal to $\cos \alpha$. The condition $\cos \alpha = 1/2$ implies that the angle α is equal to $60°$. The lifeguard does not need to measure any distances — only the angle α.

A faster track

Galileo, in some of his experiments, rolled a ball down an inclined plane and then allowed the ball to continue to roll on a horizontal plane, for example, the path OQP in Figure 12.4(b). We idealize the situation in two ways. (1) We replace the ball with a mass particle that slides without friction. This avoids any consideration of the rotation of the ball. (2) The ball experiences a bump on encountering the change of direction at Q. However, we assume that no energy is lost at this point — that the direction, but not the magnitude, of the velocity changes instantaneously at point Q. In practice, this assumption can be achieved only approximately.

Perhaps a straight track with a horizontal extension is a promising third entry in the roller-coaster race, discussed on page 210, where it was shown that, in Figure 11.10, the straight track achieves a faster descent than a certain curved track. Later in this chapter, we will see that a curved track in the shape of a cycloid achieves the least possible time of descent.

Problem 12.4 (inclined plane with a horizontal extension). In Figure 12.4, starting from rest at point O, can the point Q be chosen so that a particle descends on path OQP in less time than on a straight ramp connecting O and P? If so, what is the optimal location of the point Q?

"A straight line is the shortest distance between two points," but this does not help to answer the above question, which asks for the shortest *time*, not distance.

From Problem 11.2, the mean speed of the particle sliding down the inclined plane OQ is exactly half of the speed on the remaining path QP. Therefore, Problem 12.4 is essentially the same as the lifeguard problem, Problem 12.3. According to the solution of the lifeguard problem, if the angle OPR is at least $60°$, as in Figure 12.4(a), then, for the fastest descent, the inclined plane should coincide with the line segment OP. Otherwise,

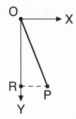

(a) Inclined plane. Angle OPR is at least 60°.

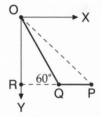

(b) Inclined plane with a horizontal extension. Angle OPR is less than 60°.

Figure 12.4. Descent from O to P on an inclined plane without or with a horizontal extension. (a) If angle OPR is 60° or steeper, a horizontal extension provides no benefit; descent, on the inclined plane OP is faster. (b) If angle OPR is less than 60°, descent on OQP, a 60° inclined plane with a horizontal extension, is faster.

as in Figure 12.4(b), place the inclined plane OQ so that the angle OQR is equal to 60°.

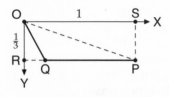

Figure 12.5. Comparing times of descent: OP and OQP.

We illustrate the above construction with a numerical example, the race discussed on page 210. Figure 12.5 shows the straight dashed path OP as in Figure 11.10—a descent from the origin O to the point P with coordinates $(1, 1/3)$. Using *short chains* (64 foot units) as the measure of distance, it was found on page 210 that the time of descent on the path OP was 3.65148 seconds. We now compute the time of descent along the path OQP, where Q is chosen so that the angle OQR is 60°. The length of RQ is equal to $\frac{1}{3}\cot 60° \approx 0.19245$, and the length of OQ is twice that amount: 0.38490. The length of QP is $1 - 0.19245 = 0.80755$. The constant speed v_1 on QP is $\sqrt{1/3} = 0.57735$, and the mean speed v_2 on OQ is $\frac{1}{2}\sqrt{1/3} = 0.28868$. The total time is given by

$$\frac{\overline{QP}}{v_1} + \frac{\overline{OQ}}{v_2} \approx \frac{0.80755}{0.57735} + \frac{0.38490}{0.28868} \approx 2.73205 \text{ seconds}$$

This shows that the inclined plane with the horizontal extension, OQP, wins the race by $3.65147 - 2.73205 \approx 0.9$ seconds—a huge margin, about 34%. The inclined plane with the horizontal extension has the advantage that the particle enjoys the maximum speed along most of its track. On the other hand, the mean speed along the straight ramp is only half of the maximum speed. However, we will see that the particle can make the descent along a certain curved track (see Figure 12.12) in even less time.

(a) Angle ABC is greater than 120°. (b) Every angle of triangle ABC is less than 120°.

Figure 12.6. The three-towns problem. Design a system of roads that joins the three towns, A, B, and C, such that the total length of the roads is as small as possible.

A road-building project for three towns

Problem 12.5 (the three towns). Three towns, A, B, and C, shown in Figure 12.6, are located in a flat region. It has been decided to connect the towns by roads. To minimize cost, design the project so that the total length of the roads is as small as possible.

This is a problem that Fermat gave to Evangelista Torricelli (1608–47), a student of Galileo. Torricelli gave several enormously clever geometric solutions of this problem.[3] We will see that a special case of this problem is equivalent to the lifeguard and inclined plane problems, Problems 12.3 and 12.4.

The general solution is described below. There are two cases, (a) and (b), to consider:

(a) If one of the angles of the triangle is at least 120°, then the optimal system of roads consists of the two sides of the triangle that form an obtuse angle, as shown in Figure 12.6(a).

(b) Otherwise, as shown in Figure 12.6(b), connect each of the three vertices, A, B, and C, to a point P such that the angles APB, BPC, and CPA are each equal to 120°.

We will prove the following special case of the problem of the three towns, Problem 12.5:

Problem 12.6. Suppose that the points A, B, and C are the vertices of an isosceles triangle with apex at C, as shown in Figure 12.7. Find a point P on the altitude OC of the triangle such that the sum of the lengths AP, BP, and CP is as small as possible.

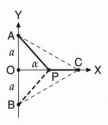

Figure 12.7. Special case of the problem of the three towns.

In Figure 12.7, the isosceles triangle ABC is shown "on its side" with the apex C to the right instead of on the top in order to emphasize the similarity between the path CPA and the path PQO in Figures 12.2 and 12.4.

Problem 12.6 can be restated without reference to vertex B as follows:

Problem 12.7. Find a point P on the line OC such that *twice* the length AP plus the length CP is as small as possible.

Stated in this form, we see that the problem is essentially the same as Problems 12.3 and 12.4. Thus we have *proved* that the solutions (a) and (b), described above, apply to this special case of the three-towns problem. That is, if the angle ACB is at least 120°, then the point P should coincide with point C; otherwise the point P should be chosen so that the angle α is equal to 60°.

The next section discusses different methods for solving extremum problems, applied, in particular, to Problem 12.2.

Inequalities

Was it really necessary to use calculus to solve Problem 12.2 on page 213? No, we will see here a solution using quite different methods—the *method of inequalities*.

Inequalities are used abundantly in mathematical analysis to make estimates. An inequality is an algebraic expression involving $<, >, \leq,$ or $\geq$. The study of inequalities that are *best possible* in some sense is an important topic in the general *theory of inequalities*. We will see that it is possible to solve an extremum problem by finding instances of a general inequality in which "$\geq$" can be replaced by "$=$."

The following solution of Problem 12.2 is *elementary* because it does not use calculus, but *difficult* because it requires us to guess the solution and then use an algebraic trick to prove that our guess is correct. In mathematics, elementary does not always mean easy. By comparison, the preceding solution of Problem 12.2 is routine and straightforward, requiring only the basic concepts of calculus.

Solution (Problem 12.2, second method). This solution is based on the fact that the square of a real number cannot be negative.

By examining Figure 12.1, it is reasonable to guess that the maximum of $A(x) = 2qx - x^2$ is achieved at $x = q$ and that the maximum value is q^2. The proof begins by subtracting $A(x)$ from the supposed maximum value q^2:

$$q^2 - (2qx - x^2) = q^2 - 2qx + x^2 \tag{12.4}$$

We must show (1) that the expression (12.4) is nonnegative and (2) that it is zero only if $x = q$. Both of these assertions follow from the fact that (12.4) is equal to a perfect square, as follows:[4]

$$q^2 - 2qx + x^2 = (q - x)^2$$

The square of a real number cannot be negative. This obvious fact is the basis for a number of conclusions that are not so immediately obvious. The next two sections are concerned with consequences of this fact: the *inequality of the arithmetic and geometric means* and *Cauchy's inequality*.[5]

The inequality of the arithmetic and geometric means

Most everyone is familiar with the idea of a mean or average of two or more quantities. The arithmetic mean is most familiar—add the quantities and divide by their number. It is traditional that examination scores, for example, are averaged in this way. However, there are other concepts of mean or average. For example, the geometric mean—if there are n nonnegative quantities, the geometric mean is defied as the nth root of their product. That is, the geometric mean of the nonnegative quantities $x_1, x_2, \ldots, x_n$ is defined to be $\sqrt[n]{x_1 x_2 \ldots x_n}$.

The geometric mean is a more suitable average than the arithmetic for data that tends to grow or decline geometrically, such as stock market prices. Benford's law tends to hold for data of this type:

> **Benford's law.** Before the advent of the computer, slide rules and tables of logarithms were used heavily for a variety of calculations. In 1938, Frank Benford, a physicist at General Electric, observed that the first few pages of tables of logarithms were more soiled than the rest. Why did people more often look up the logarithms of numbers with 1 as the leading digit? Benford found an explanation. The entries in most tables of numerical data exhibit a greater incidence than expected—about 30%—of numbers with 1 as the most significant digit. The observed rate of 30% becomes less mysterious when we notice that numbers with leading digit equal to 1 account for about 30% of the length of a slide rule. (See Figure 12.8.) This phenomenon, although it is known as *Benford's law*, was first discovered in 1881 by the American astronomer Simon Newcomb. Benford's law is used to detect falsified data because cheaters generally use 1 as an initial digit at a rate lower than 30%. Most infinite geometric progressions obey Benford's law.

Figure 12.8. A slide rule is a logarithmic scale. Numbers with leading digit equal to 1 account for 30% of the length of a slide rule because $\log_{10} 2 = 0.30103$.

Theorem 12.2 (inequality of arithmetic and geometric means). *Let the numbers x_i, $i = 1, \ldots, n$, be nonnegative. Then the arithmetic mean, $(x_1 + x_2 + \cdots + x_n)/n$, is not less that the geometric mean, $\sqrt[n]{x_1 x_2 \ldots x_n}$. Moreover, the two means are equal only if $x_1 = x_2 = \cdots = x_n$.*

Proof. Although this theorem is true for all natural numbers n, we prove it here only for $n = 2$ and $n = 4$.

n = 2 : We wish to show $(x_1 + x_2)/2 \geq \sqrt{x_1 x_2}$, or, equivalently,

$$(x_1 + x_2)^2 \geq 4 x_1 x_2$$

This follows from the following algebraic calculation:

$$(x_1 + x_2)^2 - 4 x_1 x_2 = x_1^2 + 2 x_1 x_2 - 4 x_1 x_2 = x_1^2 - 2 x_1 x_2 + x_2^2 = (x_1 - x_2)^2$$

The last term is nonnegative; moreover, it is zero only if $x_1 = x_2$.

n = 4 : We wish to show

$$\tfrac{1}{4}(x_1 + x_2 + x_3 + x_4) \geq \sqrt[4]{x_1 x_2 x_3 x_4}$$

This is a double application of the preceding case ($n = 2$). In fact, we can write:

$$\tfrac{1}{4}(x_1 + x_2 + x_3 + x_4) = \tfrac{1}{2}\left(\tfrac{1}{2}(x_1 + x_2) + \tfrac{1}{2}(x_3 + x_4)\right)$$

$$\geq \tfrac{1}{2}\left(\sqrt{x_1 x_2} + \sqrt{x_3 x_4}\right) \geq \sqrt{\sqrt{x_1 x_2}\sqrt{x_3 x_4}} = \sqrt[4]{x_1 x_2 x_3 x_4}$$

Equality holds throughout only if $x_1 = x_2 = x_3 = x_4$. □

The following is an easy corollary of Theorem 12.2.

Corollary 12.2.1. *A positive real number x plus its reciprocal $1/x$ is at least 2. Equality holds only if $x = 1$.*

Proof. For any $x > 0$, Theorem 12.2 implies

$$\tfrac{1}{2}\left(x + \frac{1}{x}\right) \geq \sqrt{x \cdot \frac{1}{x}} = 1$$

Multiply both sides of the equation by 2 to obtain the claimed assertion. Equality holds only if $x = 1/x$, that is, only if $x = 1$. □

Cauchy's inequality

The following inequality is named after the French mathematician Augustin-Louis Cauchy (1789–1857).

Theorem 12.3 (Cauchy's inequality—special case[6]). *Let a, b, c and d be arbitrary real numbers. Then we have*

$$(ab + cd)^2 \leq (a^2 + c^2)(b^2 + d^2) \tag{12.5}$$

Equality holds only if the proportion a : b = c : d is true. In other words, equality holds only if ad = cb.

Proof. It is sufficient to show that the right side minus the left side of (12.5) is a perfect square—as shown by the following algebraic calculation:

$$(a^2 + c^2)(b^2 + d^2) - (ab + cd)^2$$
$$= (a^2b^2 + a^2d^2 + c^2b^2 + c^2d^2) - (a^2b^2 + 2abcd + c^2d^2)$$
$$= a^2d^2 + c^2b^2 - 2abcd = (ad - cb)^2 \qquad \square$$

Cauchy's inequality and the inequality of the arithmetic and geometric means are *sharp* inequalities. This means that the quantities involved can be chosen so that the inequality sign, $\geq$ or $\leq$, can be replaced by equality. Theorems 12.2 and 12.3 state explicitly conditions under which equality holds. In this sense, these inequalities, in fact, sharp inequalities in general, are the *best possible.*

These two inequalities will be used in the next section in the proof of an extremal problem with an interesting history.

The Brachistochrone

To this point, we have considered only extremal problems that can be solved by adjusting the value of a single variable. For example, Problem 12.2, to find the rectangle of fixed perimeter having the largest area, is such a problem. It might seem that in this problem *two* variables must be determined; however, specifying the width determines the length because the perimeter is fixed. A problem of this sort is said to have *one degree of freedom.* On the other hand, to find a box with fixed surface area having the greatest volume is a problem with *two* degrees of freedom.

There is a deeper class of extremal problems having *infinite* degrees of freedom. For example, to find a closed curve with fixed perimeter that encloses the largest possible area is a problem with infinite degrees of freedom. It not an easy matter to prove that the extremal curve is a circle.

Extremal problems with infinite degrees of freedom are called *variational problems.* The technique for solving such problems is called the *calculus of variations.* The principles mentioned on the first page of this chapter, for example, *Fermat's principle* of least time and the *principle of least action,* are variation principles.

Problem 12.4, the problem of the quickest descent down an inclined plane with a horizontal extension, is a problem with one degree of freedom. However, this problem has a natural extension to a variational problem known as the *brachistochrone* problem.

Problem 12.8 (the brachistochrone). Suppose, as shown in Figure 12.9, that a particle, starting from rest at point O, descends under gravity without friction on a curved track connected to a lower point P. Find a curve connecting O and P such that the time of descent is minimum.

Henceforth, we will assume that all the competing tracks are contained in a fixed vertical plane that contains the points O and P. In fact, it can be shown that this restriction does not change the minimum time of descent.

The curve that solves the brachistochrone problem is the *cycloid,* shown in Figure 12.9. In that figure, the particle starts its descent at point O and terminates at point P. The cycloid is the plane curve traced by point R on the circumference of a circle as it rolls upside down without slipping on a straight line, the *x*-axis. The diameter of the generating circle can be chosen so that the tracing point R, which is initially at O, eventually passes through point P. The cycloid was mentioned previously as an example of a mechanically generated curve.[7]

The last brick of the proof of Problem 12.8 will be put in place on page 238. In the meantime, we will find many preliminary results that are interesting in themselves. This "last brick" is the following proposition. It can be proved now, even before we have established that the brachistochrone is a cycloid. Informally speaking, this proposition says that if

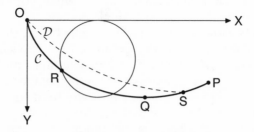

Figure 12.9. The cycloid C is the brachistochrone relative to points O and P. Point R is the tracing point on the rolling circle. Point Q is the lowest point on the cycloid, corresponding to a half rotation of the circle. For convenience in later calculations, the *y*-axis points downward. The point S is an arbitrary point on the cycloid, and the dashed curve D is an arbitrary alternate curve connecting the initial point O to point S.

a curve is a brachistochrone, then certain subarcs of the curve *inherit* the brachistochrone property.

Proposition 12.1. *Suppose that the curve C in Figure 12.9 is the brachistochrone relative to points O and P; and suppose that S is an arbitrary point on C. Then the arc OS of the curve C is the brachistochrone relative to the points O and S.*

Proof. We use the method of proof by contradiction. Suppose there were a path $\mathcal{D}$ of faster descent from O to S. The idea of the proof is to show that this assumption leads to a contradiction. In fact, we show that, by joining the curve $\mathcal{D}$ with the arc SP, we obtain a curve $\mathcal{E}$ that gives a more rapid descent from O to P than $\mathcal{C}$ — contradicting the fact that $\mathcal{C}$ is the brachistochrone relative to O and P.

The time to travel the arc SP depends only on the speed of the particle when it reaches S. Moreover, equation (11.14) on page 209 shows that the speed of the particle depends only on its depth. Thus, the particle's speed when it arrives at point S is the same whether it travels on $\mathcal{D}$ or on the arc OS of the cycloid $\mathcal{C}$. Thus, our assumption concerning the path $\mathcal{D}$ leads to the contradiction that path $\mathcal{E}$ achieves a faster descent from O to P than path $\mathcal{C}$. □

An experimental demonstration of the brachistochrone can be achieved by sliding a bead down a wire in the shape of a cycloid. Simultaneously, slide another bead down a straight wire connecting the same end points. The bead on the cycloid wins the race — at least, it should win. A certain distinguished professor — I forget his name — was noted for performing this experiment for his classes. A colleague once asked him, "I tried your demonstration, but when I did it, the bead on the straight wire won the race — contrary to the predicted outcome. How do you avoid this embarrassment?" Without hesitation he replied, "I grease the cycloid."

If the demonstration fails, it is not because of a flaw in the theory. It is because *friction* tends to spoil the experiment. The theory requires that there is no friction whatever. The demonstration succeeds if friction is sufficiently small.

Galileo was the first to formulate the brachistochrone problem. Galileo was aware as early as 1638 that the straight line did not achieve the minimum time of descent for his inclined plane experiments. He gave an erroneous demonstration that an arc of a circle is the brachistochrone.

The Contest

In 1696, the Swiss mathematician Johann Bernoulli (1667–1748) published the brachistochrone problem in the form of a challenge to the entire mathematical community:

The Brachistochrone Challenge

I, Johann Bernoulli, greet the most clever mathematicians in the world. Nothing is more attractive to intelligent people than an honest, challenging problem whose possible solution will bestow fame and remain as a lasting monument. Following the example set by Pascal, Fermat, and so on, I hope to earn the gratitude of the entire scientific community by placing before the finest mathematicians of our time a problem which will test their methods and the strength of their intellect. If someone communicates to me the solution of the proposed problem, I shall then publicly declare him worthy of praise.

Today, Bernoulli's challenge seems a display of quaint bravado. I suspect that today's scientists have the same motivations, although they tend to mute their public displays of feeling.

Several distinguished mathematicians responded to Bernoulli's challenge and submitted solutions to the brachistochrone problem. Newton submitted his solution anonymously. Bernoulli guessed Newton's identity, saying, "I recognize the lion by his paw." Other mathematicians who submitted solutions were Johann's older brother Jakob Bernoulli (1654–1705), Gottfried Leibniz, and the French mathematician Guillaume François Antoine de l'Hôpital (1661–1704).

The brachistochrone problem has application to the sport of skiing because skis have a very low coefficient of friction on snow.[8] In a downhill race, a skier may find that a detour that approximates a cycloid is faster than a shorter alternate straight-line route.

The solutions elicited by Johann Bernoulli's challenge, and Bernoulli's own solution, showed great cleverness. Johann Bernoulli arrived at a solution by translating the brachistochrone into the problem of the path of a ray of light passing through a suitable nonhomogeneous medium. Today, the brachistochrone problem is often used as an introduction to the general tools of the calculus of variations. Such a treatment is beyond the scope of this book. Moreover, the standard treatment of the brachistochrone problem using the calculus of variations generally suffers from a subtle flaw, *the assumption that there exists a solution.* The solutions by Bernoulli and the others also suffer from this defect. For simpler extremal problems, this difficulty was discussed on page 215.

In the following pages, I would like to present my own solution (Benson (1969)) to the brachistochrone problem. This solution has two advantages: (1) It uses nothing more advanced than basic calculus, and (2) it answers the existence problem mentioned above.

First, we need to examine a geometrical construction of a tangent line to a cycloid.

The geometry of the cycloid

In this section we will find the slope of a tangent to a cycloid curve. In Figure 12.10, point R with coordinates (x, y) is the point on the rolling circle of diameter M that traces the cycloid C.

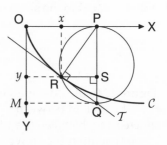

Problem 12.9. Show that the slope of the tangent to the cycloid C in Figure 12.10 at an arbitrary point R is equal to

$$\pm\sqrt{\frac{M}{y} - 1}$$

Figure 12.10. Construction of a tangent line T to the cycloid C.

Solution. Point P is the point of contact between the x-axis and the cycloid C, and PQ is a diameter of the circle ($\overline{PQ} = M$).

Point P is the instantaneous center of rotation of the rolling circle. Consequently, the line PR is normal (i.e., perpendicular) to the cycloid at R. Angle PRQ is a right angle because it is inscribed in a semicircle. Since the line T through points RQ is perpendicular to the normal line RP, T is a tangent line to the cycloid with point of tangency at R.

Draw the line RS perpendicular to PQ. The slope of the tangent T is equal to the quotient $\overline{SQ}/\overline{RS}$. We will solve the problem by expressing this fraction as a function of y and M. It is easy to express $\overline{SQ}$ in this way; in fact, we have $\overline{SQ} = M - y$.

It is slightly more complicated to express $\overline{RS}$ as a function of y and M. There is a standard theorem of plane geometry that asserts

$$\overline{RS}^2 = \overline{PS} \cdot \overline{SQ} \tag{12.6}$$

This fact can be derived from the fact that triangle SRP is similar to triangle SQR. From equation (12.6), we obtain

$$\overline{RS}^2 = y(M - y)$$

Finally, we obtain the slope m of T in the required form:

$$m = \overline{SQ}/\overline{RS} = \pm\frac{M - y}{\sqrt{y(M - y)}} = \pm\sqrt{\frac{M}{y} - 1}$$

The positive sign applies for the point R as shown in Figure 12.10. The negative sign applies, for example, at point P in Figure 12.9.

The cycloid curve in Figure 12.9 can be considered the graph of a function $y = f(x)$—let us call it the *cycloid function*. This function has a derivative that is equal to the slope of the tangent of the cycloid:

$$\frac{dy}{dx} = \pm\sqrt{\frac{M}{y} - 1} \qquad (12.7)$$

In Figure 12.9, the positive sign applies where $f(x)$ is nondecreasing, the part of the cycloid between O and Q; and the negative sign where $f(x)$ is nonincreasing, between Q and P.

From equation (12.7), we can find an even simpler expression for the derivative of arc length s with respect to x. In fact, for the cycloid function, we have, using equations (11.11) and (12.7),

$$\frac{ds}{dx} = \sqrt{1 + y'^2} = \sqrt{1 + \left(\frac{M}{y} - 1\right)} = \sqrt{\frac{M}{y}} \qquad (12.8)$$

A differential equation

An equation, like (12.7), that involves the derivative of a function is called a *differential equation*. We found above that the cycloid function satisfies the differential equation (12.7). We will find below that a solution of a modified brachistochrone problem also satisfies the differential equation (12.7).

To *solve* a differential equation means to solve the inverse problem: Given a differential equation, like (12.7), find *all* functions that satisfy that equation.

The preceding sentence needs some qualification. How big an "all" is necessary? Answer: A universe of curves, of functions, that is reasonably simple but big enough for the intended discussion. The following definition makes use of a coordinate system in which the y-axis points *downward*.

Definition 12.1. Let x_0 be a fixed nonnegative number. A *smooth downward ramp* on the interval $I = (0 \leq x \leq x_0)$ is the graph of a function $y = f(x)$, defined all for x in I, satisfying the following properties:

1. The function $y = f(x)$ is smooth. That is, the derivative $f'(x)$ exists and is continuous for all x in the interval $0 < x < x_0$. Moreover, $f(x)$ is continuous in I.
2. The function $y = f(x)$ is nondecreasing. That is, $f'(x) \geq 0$ for all x in the interval $0 < x < x_0$.

A smooth *downward* ramp corresponds to a *nondecreasing* function $y(x)$ because the y-axis points *downward*, as in Figure 12.11.

A *smooth upward ramp* is defined by making the following alterations in item 2 of the above definition: Replace "nondecreasing" with "nonincreasing" and "$f'(x) \geq 0$" with "$f'(x) \leq 0$."

The curve $\mathcal{D}$ in Figure 12.11 is an example of a graph of a function that is a smooth downward ramp. Note that it is impossible for a smooth downward ramp proceeding from O to P to drop below the elevation of the terminal point P. Like curve $\mathcal{C}$ in Figure 12.11, it is possible for a smooth downward ramp to include a horizontal segment. In fact, once the ramp drops to the elevation of the terminal point P, it must proceed along a horizontal line segment. For example, in Figure 12.11, the ramp $\mathcal{C}$ contains the line segment RP.

One might be tempted to say that if a smooth downward ramp satisfies the differential equation (12.7), then the graph of the function must be a cycloid, but this would be jumping to a false conclusion. In fact, the constant function $y = M$ satisfies equation (12.7), but its graph is a horizontal straight line, not a cycloid.

There is a further complication: The right side of equation (12.7), the derivative of the cycloid function, has a difficulty at the initial point O; in fact, the term M/y in equation (12.7) is undefined at O because, at that point, y is equal to zero, and division by zero is not allowed. However, corresponding to the geometrical fact that the cycloid has a vertical tangent at O, the derivative of the cycloid function tends to $+\infty$ as x tends to 0 through positive values. Notice that item 2 above does not require that the derivative exist for $x = 0$. The following fact characterizes a smooth downward ramp $y = f(x)$ that satisfies equation (12.7):

Fact 12.1. *As in Figure 12.11, let $\mathcal{D}$ be the graph of a smooth downward ramp $y = f(x)$ connecting the points O : $(0,0)$ and P : (x_0, M), and let the following conditions be satisfied:*

1. x_0 is at least equal to the circumference x_1 of a semicircle of diameter M:

$$x_0 \geq \tfrac{\pi}{2} M$$

2. For all x in the interval $0 < x \leq x_0$, the derivative $dy/dx = f'(x)$ satisfies the equation

$$\frac{dy}{dx} = \sqrt{\frac{M}{y} - 1} \qquad (12.9)$$

Then the graph of $y = f(x)$ is the curve $\mathcal{C}$ in Figure 12.11: half of an arch of the cycloid OQ, followed (if $x_0 > x_1$) by the horizontal line segment QP : $y = M$ for all x in the interval $x_1 < x \leq x_0$.

The restricted brachistochrone

This section continues the search, begun on page 206, for a roller coaster with a more rapid descent. We have seen (Problem 12.4 and Figure 12.4) that sometimes a 60° ramp with a horizontal extension shortens the time of descent compared to a straight-line ramp. We consider below an extension of this problem: How should we design a curved roller-coaster track, for example, the dashed curve $\mathcal{D}$ in Figure 12.11, to achieve the descent from points O to P in the least time? We depart from the standard brachistochrone problem by assuming that the elevation of the track is nonincreasing — that the track never proceeds from a lower to a higher elevation. This assumption of nonincreasing elevation is satisfied, for example, by the inclined plane with a horizontal extension in Figure 12.4. This assumption excludes the cycloid curve in Figure 12.9 because on the arc QP the elevation of the sliding particle increases. Using the roller-coaster image, the point P is at ground level, and the track is not permitted to go below ground. We will see that in Figure 12.11 the curve $\mathcal{C}$, the half arch of a cycloid with a horizontal extension, achieves the descent in least time. The roller-coaster passengers will complete the ride in less time — although they will miss the thrilling spine-crushing bump at point R in Figure 12.4.

The following is a more precise mathematical statement of the problem.

Problem 12.10. Find a smooth downward ramp $\mathcal{D}$, the graph of a function $y = f(x)$ (defined for x in the interval $0 \le x \le x_0$) such that a mass particle starting at rest from point O in Figure 12.11 descends under gravity without friction on the ramp $\mathcal{D}$ from point O to the point P in the least possible time. The smooth downward ramp $\mathcal{D}$ must carry the particle forward a fixed distance x_0 and downward a distance M. In other words:

1. Point O has coordinates $(0,0)$; that is, $y(0) = 0$.
2. Point P has coordinates (x_0, M); that is, $y(x_0) = M$.

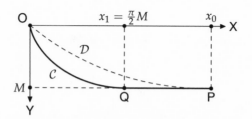

Figure 12.11. The graph $\mathcal{C}$ of $y = f(x)$, the solution of the differential equation (12.9), is a half arch of a cycloid joined to a horizontal line segment. The dashed curve $\mathcal{D}$ represents an arbitrary smooth downward ramp connecting O and P. It is shown below that, of all smooth downward ramps $\mathcal{D}$, the curve $\mathcal{C}$ — the cycloid $\mathcal{C}$ with horizontal extension — achieves the descent from O to P in minimum time.

In order to proceed with this problem, we need to find the time of descent. First, we must find the speed of the particle. On page 209, we found that if we measure distances in *short chains* (64 foot units), and if we assume the particle at rest at $y = 0$, the speed v is given by

$$v = \sqrt{y} \text{ short chains/second} \qquad (12.10)$$

In equation (11.19) we found that the time of descent T is given by the integral

$$T = \int_0^{x_0} \frac{1}{v} \frac{ds}{dx} dx$$

In this integral, v is the speed and s is the arc length on the curve $\mathcal{D}$. Using equation (11.11) on page 207 and equation (12.10) above, we find the following expression for the time of descent:[9]

$$T = \int_0^{x_0} \frac{\sqrt{1 + y'^2}}{\sqrt{y}} dx \qquad (12.11)$$

The following proposition is the key to the solution of Problem 12.10. The proof of this proposition requires the use of inequalities discussed on pages 220–223.

Proposition 12.2. *Let $\mathcal{D} : y = f(x)$ be a smooth downward ramp connecting from the initial point O : $(0,0)$ to the final point P : (x_0, M). (For example, the dashed curve $\mathcal{D}$ in Figure 12.11.) Then, assuming that distances, for example, x_0 and M, are measured in* short chains *(64 foot units), the time of descent T satisfies the inequality*

$$T \geq \frac{x_0}{\sqrt{M}} + \tfrac{1}{2}\pi\sqrt{M} \qquad (12.12)$$

Equality holds in (12.12) if and only if the function $y = f(x)$ satisfies, for all x in the interval $0 < x < x_0$, the differential equation:

$$y' = \sqrt{\frac{M}{y} - 1} \qquad (12.13)$$

Proof. Recall that Cauchy's inequality, Theorem 12.3, asserts that for any real numbers a, b, c, and d, the following inequality holds:

$$(ab + cd)^2 \leq (a^2 + c^2)(b^2 + d^2) \qquad (12.14)$$

Apply this inequality with the following substitutions:

$$a = 1 \quad b = \sqrt{\frac{1}{M}} \quad c = y' \quad d = \sqrt{\frac{1}{y} - \frac{1}{M}} \qquad (12.15)$$

Note that, in the definition of d, the quantity under the square root sign is nonnegative because we have assumed $f(x) \leq M$; that is, $y \leq M$. We now substitute the values in (12.15) for a, b, c, and d in inequality (12.14), obtaining

$$(a^2 + c^2)(b^2 + d^2) = (1 + y'^2)\left(\frac{1}{M} + \left(\frac{1}{y} - \frac{1}{M}\right)\right) = \frac{1 + y'^2}{y}$$

$$ab + cd = \frac{1}{\sqrt{M}} + y'\sqrt{\frac{1}{y} - \frac{1}{M}}$$

Now apply Cauchy's inequality to the expression (11.19) for the descent time T, obtaining

$$\int_0^{x_0} \frac{\sqrt{1 + y'^2}}{\sqrt{y}}\, dx \geq \frac{1}{\sqrt{M}} \int_0^{x_0} dx + \int_0^{x_0} \sqrt{\frac{1}{y} - \frac{1}{M}}\, y'\, dx \qquad (12.16)$$

It can be shown that the second integral above in inequality (12.16) is equal to $\frac{1}{2}\pi\sqrt{M}$. Although it is beyond the scope of this book to evaluate this integral, it would be a standard problem in a course in integral calculus. If we accept this value, $\frac{1}{2}\pi\sqrt{M}$, then inequality (12.16) is equivalent to the claimed inequality (12.12).

It remains to show that equality holds in inequality (12.12) if and only if the function $y = f(x)$ satisfies the differential equation (12.13). Recall that inequality (12.16) was obtained through an application of Cauchy's inequality (12.14) with the substitutions (12.15). According to Theorem 12.3, equality holds in Cauchy's inequality if and only if $ad = bc$, that is, if and only if

$$\sqrt{\frac{1}{y} - \frac{1}{M}} = \frac{1}{\sqrt{M}} y'$$

The above equation is equivalent to the claimed differential equation (12.13). $\square$

Notice that the right side of inequality (12.12) is a *lower bound* on T that does not depend on the choice of the function $y = f(x)$. If we can show that inequality (12.12) is sharp, in other words, if we can find a smooth downward ramp that achieves this lower bound, we will have solved the restricted brachistochrone problem. Indeed, Fact 12.1 asserts that if the distance x_0 is sufficiently large relative to M—specifically, if $x_0 \geq \frac{\pi}{2}M$—then a half arch of a cycloid followed by a portion of the horizontal line $y = M$, the curve C in Figure 12.11, is the one and only smooth downward ramp that solves the differential equation (12.13). Therefore, the curve C is the one and only solution of this restricted brachistochrone problem.

Notice that the question, raised on page 226, regarding the *existence* of a solution has been answered. We have identified the curve of minimum descent without previously assuming its existence.

The number $x_1 = \frac{\pi}{2}M$ is half of the circumference of a circle of diameter M, the x-coordinate of the point on a cycloid obtained by rolling the generating circle half a turn, the lowest point on the cycloid — for example, in Figure 12.11, the point Q.

The following proposition summarizes the above facts.

Proposition 12.3. *If the distance x_0 is sufficiently large — specifically, if $x_0 \geq \frac{\pi}{2}M$ — inequality (12.12) is sharp. In particular, equality is achieved if the smooth downward ramp is, like the curve C in Figure 12.11, a downward half arch of a cycloid (OQ), followed, if $x_0 > \frac{\pi}{2}M$, by a segment of the horizontal line $y = M$ (QP). In other words, the curve C solves the restricted brachistochrone problem, and the corresponding minimum time T is given by*

$$T = \frac{x_0}{\sqrt{M}} + \tfrac{1}{2}\pi\sqrt{M}$$

A loose end

Proposition 12.3 solves an extension of Problem 12.4, illustrated in Figure 12.4. However, something is missing. Figure 12.11, which illustrates Problem 12.3, seems analogous to Figure 12.4(b), but there does not seem to be any analogy to Figure 12.4(a), where the inclined plane is so steep that no horizontal extension is needed. This is reasonable because Proposition 12.3 requires that the drop from initial point O to the final point P is not too steep; that is the meaning of the assumption $x_0 \geq \frac{\pi}{2}M$. This loose end will be tied down on page 239 when we consider the possibility of a steep drop from O to P, corresponding to the inequality $x_0 < \frac{\pi}{2}M$, analogous to Figure 12.4(a).

As we have seen above, inequality (12.12) is sharp if $x_0 \geq \frac{\pi}{2}M$. On the other hand, if x_0 is smaller than this bound, that is, if $x_0 < \frac{\pi}{2}M$, then inequality (12.12) is still correct, but it is not sharp. In fact, this inequality gives an estimate that is too small.

For example, consider the case of free vertical fall. That is, suppose that x_0 is equal to 0. The time for the particle to fall directly downward to the depth M is $2\sqrt{M}$ seconds. This can be seen by using the fact, stated in Problem 11.2, that the time of descent on a straight path is equal to the length of the path, M, divided by the mean speed $\frac{1}{2}\sqrt{M}$. That is, the time of descent T is given by

$$T = \frac{M}{\frac{1}{2}\sqrt{M}} = 2\sqrt{M}$$

Common sense urges, and it can be shown mathematically, that no ramp can achieve a faster descent to the point at depth M directly below the starting point. However, since $x_0 = 0$, the right side of inequality (12.12) is equal to $\frac{1}{2}\pi\sqrt{M} \approx 1.57080\sqrt{M}$ seconds. This lower bound for the time of descent is too small because we know that the fastest time of descent is actually equal to $2\sqrt{M}$ seconds. Thus, in case $x_0 = 0$, we see that the lower bound given by the right side of inequality (12.12) fails to be sharp.

Two more entries, silver and gold

Formula (12.12) can be used to compute the time of descent for another contestant in the race discussed on pages 208–211 and 217–218. This path is shown in Figure 12.12(a)—half of an arch of a cycloid followed by a

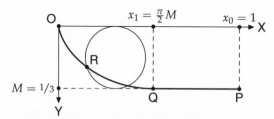

(a) Path is not permitted to drop below terminal point. Silver medal winner. Descent time = 2.63895 seconds.

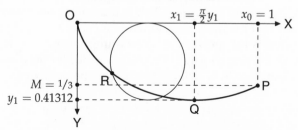

(b) Path is unrestricted. Gold medal winner. Descent time = 2.60416 seconds.

Figure 12.12. (a) Silver and (b) gold medal winners of the roller-coaster Olympics. A mass particle slides without friction under gravity starting from rest at point O, descending to point P. Figure (a) shows the path of quickest descent if the path is not permitted to drop lower than the terminal point P—half of the arch of a cycloid followed by a horizontal straight line. Figure (b) shows the quickest path if descent below the terminal point is allowed—a cycloid generated by a larger circle. For convenience of calculation, distances are measured in *short chains* (64 foot units). Point P is down $\frac{1}{3}$ unit ($21\frac{1}{3}$ feet) and forward 1 unit (64 feet) from point O. Times of descent are (a) 2.63895 seconds and (b) 2.60416 seconds.

horizonal straight line. The time of descent for this path is computed as follows. In formula 12.12, put $x_0 = 1$ and $M = 1/3$, obtaining for the time T of descent

$$T = \frac{x_0}{\sqrt{M}} + \frac{1}{2}\pi\sqrt{M} = \frac{1}{\sqrt{1/3}} + \frac{1}{2}\pi\sqrt{1/3} = 1.73205 + 0.90690 = 2.63895 \text{ secs}$$

A track that is permitted to drop below ground achieves an even better time. The cycloid shown in Figure 12.12(b) achieves a descent time of 2.60416 seconds.[10] The final results of the brachistochrone Olympics are shown in Table 12.1.

The unrestricted brachistochrone

In this section, we will see that sometimes the time of descent can be shortened — as in Figure 12.12(b) — if the path is permitted to drop below the elevation of the final point P. Where previously we allowed only smooth downward ramps, now we permit paths, like the dotted curve $\mathcal{D}$ in Figure 12.13, consisting of a smooth downward ramp — for example, the dashed curve OS in Figure 12.13 — joined to a smooth upward ramp — for example, the dashed curve SP. We will call a curve of this sort a *smooth down-up ramp*.

More precisely, we define a smooth down-up ramp to be a smooth downward ramp *possibly* joined to a smooth upward ramp. A down-up ramp, although it is permitted to rise above its deepest point, is not required to do so. In this way, the class of down-up ramps includes the class downward ramps.

We consider the special case that the final point P has the same elevation as the initial point O, as in Figure 12.13. Assuming no frictional loss, the particle, descending from O on an arbitrary smooth down-up ramp $\mathcal{D}$,

Table 12.1. Comparison of descent times of a sliding particle on various shaped downward ramps.

The Roller-Coaster Olympics Down 21 feet 4 inches, forward 64 feet		
	Time (secs)	Figure
1. Cycloid without horizontal extension	2.60416	12.12(b)
2. Cycloid with horizontal extension	2.63895	12.12(a)
3. 60° to horizontal ramp	2.73205	12.5
4. Straight ramp	3.65147	11.10 (dashed)
5. Curved ramp $y = \frac{1}{3}x\sqrt{x}$	3.73970	11.10 (solid)

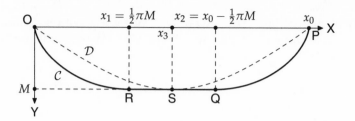

Figure 12.13.

has just enough energy to ascend back up to its initial depth at the final point P. Of course, it is no longer appropriate to call the entire journey a *descent*. We wish to find the smooth down-up ramp that completes the journey in the least time from the initial point O to the final point P.

The strategy for solving the unrestricted brachistochrone is to solve first a family of restricted brachistochrone problems — restricted by the maximum allowable depth. Then find the depth that achieves the passage from O to P in the least time.

The quantity x_0 denotes the x-coordinate of the final point P of the ramp, and M denotes the depth of the ramp. The quantities x_0 and M answer the two questions: *how far* and *how deep*.

We will find the path of least time by conducting a *tournament of two rounds*. Round 1 involves an application of Cauchy's inequality, and round 2 involves an application of the inequality of the arithmetic and geometric means.

Round 1. For each positive M, find the smooth down-up ramp of depth M that achieves the most rapid journey from point O to point P.

Round 2. The winners of round 1 compete to determine the smooth down-up ramp that achieves the most rapid journey from O to P.

Round 1: Ramps of fixed depth M

We will find that the winner of round 1 is the solid curve $\mathcal{C}$ in Figure 12.13: two half arches of a cycloid, OR and QP, connected by the horizonal line segment RQ at depth M. First we need a modification of Proposition 12.2 that applies to smooth down-up ramps.

Proposition 12.4. *Let $\mathcal{D} : y = f(x)$ be a smooth down-up ramp of depth M connecting from the initial point $O : (0,0)$ to the final point $P : (x_0, 0)$. (E.g., the dashed curve $\mathcal{D}$ in Figure 12.13. Note that the initial and final points must have the same depth.) Then, assuming that all distances, such as x_0 and M, are measured in short chains (64 foot units), the time T for a particle, initially at rest,* *to move along the ramp under gravity from* O *to* P *satisfies the inequality*

$$T \geq \frac{x_0}{\sqrt{M}} + \pi\sqrt{M} \qquad (12.17)$$

Let (x_3, M) *be an arbitrary point of depth* M *on the ramp* $\mathcal{D}$, *for example, point* S *in Figure 12.13. Equality holds in* (12.17) *if and only if the function* $y = f(x)$ *satisfies, for all* x *in the interval* $0 < x < x_0$, *the differential equation*

$$y' = \pm\sqrt{\frac{M}{y} - 1} \qquad (12.18)$$

In the above equation, "$\pm$" is taken to be "$+$" for x *in the interval* $0 < x < x_3$ *and "$-$" in the interval* $x_3 \leq x < x_0$.

Proof. Let T_1 and T_2 be, respectively, the time to traverse from $x = 0$ to $x = x_3$ (in Figure 12.13, from O to S) and from $x = x_3$ to $x = x_0$ (in Figure 12.13, from S to P).

By Proposition 12.2—Cauchy's inequality was used to establish this proposition—the time T_1 satisfies the inequality

$$T_1 \geq \frac{x_3}{\sqrt{M}} + \tfrac{1}{2}\pi\sqrt{M}$$

The time T_2 for the particle to ascend from S up to P is the same as the time for a descent from P down to S. For ascent or descent, the speed at depth y is the same, $\sqrt{y}$, but in opposite directions. Thus, from Proposition 12.2, we have the inequality

$$T_2 \geq \frac{x_0 - x_3}{\sqrt{M}} + \tfrac{1}{2}\pi\sqrt{M}$$

Putting these two inequalities together, we have, as claimed,

$$T = T_1 + T_2 \geq \frac{x_0}{\sqrt{M}} + \pi\sqrt{M}$$

The condition (12.18) follows from condition (12.13). The minus sign holds for x in the interval $x_3 \leq x < x_0$ to take account of the fact that in this interval we have $y' \leq 0$. $\qquad\square$

The following analogue of Proposition 12.3 discusses further the conditions under which inequality (12.17) is sharp.

Proposition 12.5. *If the distance* x_0 *is sufficiently large—specifically, if* $x_0 \geq$ πM— *inequality* (12.17) *is sharp. In particular, equality is achieved if the smooth down-up ramp, like the curve* C *in Figure 12.13, consists of three parts: a downward half arch of a cycloid* (OR), *joined, if* $x_0 > \pi M$, *to* RQ, *a segment of the horizontal line* $y = M$, *and, finally, joined to an upward half arch of a cycloid* (QP).

Round 2: Finding the grand prize winner

To find the depth M that minimizes the time to travel from O to P in Figure 12.13, we apply the inequality of the arithmetic and geometric means, Theorem 12.2, using $n = 2$, to the right side of inequality (12.17), obtaining

$$\frac{1}{2} \left(\frac{x_0}{\sqrt{M}} + \pi\sqrt{M} \right) \geq \sqrt{\pi x_0} \tag{12.19}$$

According to Theorem 12.2, inequality (12.19) becomes an equality if and only if

$$\frac{x_0}{\sqrt{M}} = \pi\sqrt{M} \tag{12.20}$$

that is, if and only if $x_0 = \pi M$.

Putting inequality (12.19) together with inequality (12.17), we find that the time of passage from O to P in Figure 12.13 satisfies the following chain of two inequalities:

$$T \geq \frac{x_0}{\sqrt{M}} + \pi\sqrt{M} \geq 2\sqrt{\pi x_0} \tag{12.21}$$

Inequality (12.21) is the "brachistochrone oracle." If we can understand the language it speaks, we will have answers to the questions before us.

As we have seen, the first inequality sign of (12.21) leads to the solution of the *restricted* brachistochrone problem. In fact, we saw in Proposition 12.5 that the restricted problem is solved by the down-up ramp that is shown in Figure 12.13, the ramp that gives rise to equality in the first inequality of (12.21).

The crucial and astonishing fact is that the conditions $x_0 \geq \pi M$ and $x_0 = \pi M$ for sharpness of the two inequalities in (12.21) can be satisfied *simultaneously*. The first inequality in (12.21) is sharp if $x_0 \geq \pi M$ and the ramp satisfies the differential equation (12.18). The second inequality in (12.21) is sharp if $x_0 = \pi M$. The ramp that satisfies both these conditions for equality in (12.21) is the solid curve shown in Figure 12.14 — a full arch of a cycloid with no horizontal line segment interpolated. Thus, we see that the unrestricted brachistochrone problem is solved by the smooth down-up ramp consisting of a full arch of a cycloid, as in Figure 12.14, and we have solved Problem 12.8.

Fastest descent to *anywhere*

We have solved the unrestricted brachistochrone problem in the case that the final point P has the same depth as the initial point O. However, an application of Proposition 12.1 will show that we have solved the problem *no matter where the final point is*, provided the final point is no higher than the initial point. Referring to Figure 12.14, Proposition 12.1 tells us that optimality of the cycloid among all ramps that connect O with P is inherited by any partial arch of the cycloid C, for example, the solid curve OS. In other words, the cycloidal ramp OS achieves a faster descent than any other smooth down-up ramp D connecting O to S — for example, the dotted curve D in Figure 12.14.

Figure 12.15 illustrates how the entire quarter plane to the right of and below O can be covered by the family of cycloidal arches emanating from O. An arbitrary point P to the right of O, but no higher than O, lies on exactly one cycloid of this family. This cycloid is the unique path of quickest descent from O to P. These facts are summarized in the following proposition:

Proposition 12.6. *Suppose that P is an arbitrary point to right of, and no higher than, the initial point O. Then there is exactly one cycloid arch C that has a vertical tangent at O and that passes through the point P. This cycloidal arch is the down-up ramp that achieves the most rapid descent from O to P.*

Figure 12.15 facilitates a comparison between the content of Propositions 12.6 and 12.3. Recall that Proposition 12.3 specifies the most rapid descent on a *downward* ramp. In particular, in Figure 12.15, the fastest downward ramp from O to P is half arch from OQ joined to the horizontal line segment QP. On the other hand, Proposition 12.6 specifies that the most rapid descent on a *down-up* ramp is the dashed cycloid OP.

Tying down a loose end

On page 233, we examined a gap in the analogy between Problem 12.4, *the inclined plane with horizontal extension*, and Proposition 12.3, *the restricted brachistochrone*. Now we have the tools necessary to tie down this "loose end."

In Problem 12.4, two cases were distinguished, as shown in Figures 12.4(a) and (b):

(a) If the inclined plane connecting O and P is sufficiently steep, steeper than 60°, then a horizontal extension provides no benefit.

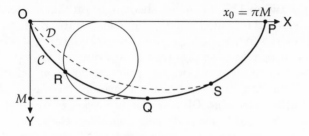

Figure 12.14. The unrestricted brachistochrone.

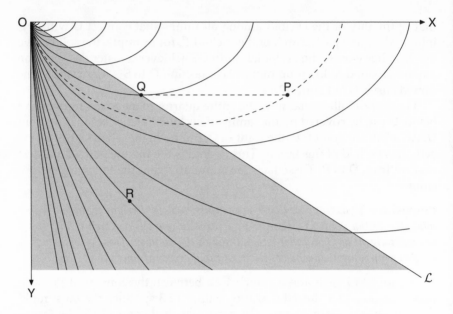

Figure 12.15. Covering a quarter plane with a family of cycloids.

(b) Otherwise a 60° inclined plane followed by a horizontal ramp is optimal.

Proposition 12.3, *the restricted brachistochrone*, which puts a limit on the steepness of the drop from O to P, provides an analogy to (b), but not to (a). However, we can now fill in this lack.

The point Q in Figure 12.15 lies on the half line $\mathcal{L}$, consisting of the points satisfying $x = \frac{\pi}{2}y$, $y > 0$. The line $\mathcal{L}$ consists of the deepest points on all of the cycloids of the family. Proposition 12.3 says that *if* P *is to the right of* $\mathcal{L}$, the *downward* ramp of fastest descent is the curve OQP, described above. The region to the left of $\mathcal{L}$ is shown shaded because Proposition 12.3—by assuming the inequality $x_0 \geq \frac{\pi}{2}M$—leaves us "in the dark" about what happens if the terminal point (e.g., R) is in the shaded region. Proposition 12.6, however, sheds light on this matter because it does not exclude the shaded region from consideration. Moreover, if the terminal point R is in the shaded region, the cycloidal arc OR is a *downward* ramp—and, also, a *down-up* ramp, because a down-up ramp, although it is permitted to rise above its deepest point, is not required to do so; in other words, the class of down-up ramps includes the class of downward ramps. *Since the cycloidal arc* OR *is optimal in the class of down-up ramps, it is also optimal in the class of downward ramps connecting* O *and* R.

The preceding remark achieves two ends: (1) It complements Proposition 12.3 by specifying the optimal descent when the drop from the initial point O to terminal point P is too steep to be covered by the proposition. (2) It completes the analogy with the part of Problem 12.4 discussed in (a) above and illustrated in Figure 12.4(a).

On the one hand, the brachistochrone problem is merely a curious puzzle, albeit one that attracted serious attention of the mathematical giants of the late seventeenth century. On the other hand, this problem signaled the beginning of the calculus of variations, a cornerstone of mathematical physics.

In this book, we have seen many other curiosities that have led to important mathematics. I hope that in these pages I have shown these two sides of mathematics: on the one hand, the puzzler, the poser, the brain-teaser; on the other, the search for understanding of the world we live in. To those who wish to continue this search, I wish enjoyable and successful further explorations.

Glossary

A

algorithm A completely specified step-by-step mathematical procedure — a "mathematical recipe," for example, long division. Computer science is largely concerned with algorithms. p. 14.

anthyphairesis The process of back-and-forth subtraction (BAFS) that defines a **ratio** in the sense of Theaetetus. The anthyphairesis of a **ratio** is equivalent to the representation of a number by a simple **continued fraction**. p. 26.

axiomatic method A process whereby all assertions of an area of discourse are derived by specified rules of inference from basic propositions called axioms or postulates referring to a small number of primitive undefined terms (for example, in Euclidean geometry, *line* and *point* are undefined). p. 74.

B

bilateral symmetry Organization of a two- or three-dimensional object (organism, art work, etc.) in which the right and left sides are mirror images of each other. p. 142.

binary system The representation of numbers as a sum of powers of 2. A number is represented in the binary system using 0 and 1 as digits. E.g., the number 11 in the binary system is:

$$8 + 2 + 1 = 1 \cdot 2^3 + 0 \cdot 2^2 + 1 \cdot 2^1 + 1 \cdot 2^0 = 1011$$

Computers represent numbers internally in the binary system. p. 8.

brachistochrone The curve that achieves the minimum time of descent. p. 224.

byte A unit of computer storage comprising eight bits of information. One byte can be used to encode one character or an integer between 0 and 255. A character can be a lower or upper case letter; or a punctuation mark, a space, or other special symbol. p. 83.

C

chain A surveyor's measure of length equal to 66 feet. In the discussion of mass particles sliding on ramps, this book uses a unit of length called a *short chain*, equal to $2g \approx 64$ feet. p. 208.

commensurable Possessing a common measure. More precisely, two magnitudes are said to be commensurable if their **ratio** is equal to the **ratio** of two **natural numbers**. p. 21.

commutative law A property of a binary operation, for example, ordinary addition or multiplication, whereby the result of applying the operation to two elements is independent of the order of application, for example, $a + b = b + a$, $ab = ba$. p. 146.

continued fraction A fraction of the form

$$a_0 + \cfrac{b_1}{a_1 + \cfrac{b_2}{a_2 + \cfrac{b_3}{a_3 + \cdots}}}$$

where a_0, a_1, ... and b_1, b_2, ... are **natural numbers**—called a *simple* **continued fraction** if $b_n = 1$, $n = 1, 2, \ldots$. There is a connection between the theory of continued fractions and (1) the Euclidean Algorithm and (2) anthyphairesis. p. 22.

coplanar Lying in the same plane. p. 75.

curvature of a curve The curvature at a point on a curve is the reciprocal of the radius of the **osculating circle**. The larger the curvature, the "tighter" the curve. p. 63.

D

difference quotient Suppose that the values of a variable y depend on the values of a variable x, and suppose y_1 and y_2 are values of the variable y corresponding to distinct values x_1 and x_2, respectively, of the variable x. Putting $k = y_2 - y_1$ and $h = x_2 - x_1$, the difference quotient is defined as k/h. For the graph of y vs. x, the difference quotient is the slope of the secant through the points (x_1, y_1) and (x_2, y_2). If y is distance and x is time, then the difference quotient is equal to the mean velocity over the time interval (x_1, x_2). p. 172.

discriminant The discriminant of an algebraic equation is a quantity depending on the coefficients of the equation whose vanishing implies that the equation has repeated roots. E.g., the discriminant of the quadratic equation $ax^2 + bx + c = 0$ is $b^2 - 4ac$. p. 172.

distributive law The distributive law of multiplication with respect to addition says that, for any numbers a, b, and c, we have $a(b + c) = ab + ac$. p. 115.

division algorithm Given two positive real numbers a and b, the division algorithm expresses a as a multiple of b plus a remainder smaller than b. More precisely, the division algorithm determines a unique nonnegative integer q, called the quotient, and a unique real number r, called the remainder, such that $0 \le r < b$ and $a = qb + r$. p. 26.

E

equimultiple Magnitudes A and B are equimultiples of a and b, respectively, if there exists a **natural number** n such that $A = na$ and $B = nb$. p. 24.

Euclidean Algorithm A procedure for finding the **greatest common divisor** of two **natural numbers**. p. 27.

F

fungible A commodity is said to be fungible if quantity, for example, area, volume, or weight, is the sole measure of value. Grain is generally a fungible commodity — one bushel is as good as another. p. 7.

G

Gaussian curvature The intrinsic curvature of a surface: the product of the two **principal curvatures**. An ellipsoid and a saddle surface have, respectively, positive and negative Gaussian curvature. p. 66.

geodesic A path that connects every sufficiently close pair of its points by a path of shortest distance. E.g., on the surface of the earth, circles of longitude are geodesics, but circles of latitude (except the equator) are not. p. 65.

greatest common divisor The greatest common divisor (GCD) of two **natural numbers** is the largest **natural number** that is a divisor of both. The Euclidean Algorithm is a method of computing the GCD of two numbers. p. 27.

group A set with an operation of multiplication satisfying the axioms listed on page 148. The set of nonzero real numbers is a group with respect to ordinary multiplication. p. 143.

I

instantaneous velocity The instantaneous velocity of a particle at an instant of time t_0 is equal to the limit of the mean velocities over time intervals containing t_0 as the length of those time intervals tends to 0. p. 179.

integer A whole number — positive, negative, or zero. p. 16.

interval An interval of real numbers consists of all numbers between a pair of given numbers called *endpoints*. If one of the endpoints is $\pm\infty$, the interval is said to be *unbounded*. An interval that contains its endpoints is said to be closed, for example, $0 \le x \le 1$. An interval that contains neither of its endpoints is said to be open, for example, $0 < x < 1$. A point of an interval that is not an endpoint is called an *interior point*. p. 191.

intrinsic geometry Also called inner geometry; a geometry based on measurements within a particular geometric universe. E.g., the radius of a circle is not meaningful in the intrinsic geometry of the circle because it cannot be measured, or even inferred indirectly, without leaving the circle. Intrinsic geometry is trivial for a circle, but more complex for a two-dimensional surface. p. 62.

irrational number A number not expressible as the quotient of two **integers**. It can be shown that $\sqrt{2}$ is irrational. p. 21.

isometry A mapping between two spaces that preserves distances. p. 150.

isoperimetric problem The classic isoperimetric problem is to find the plane figure of greatest area having fixed perimeter. The problem has been generalized in various ways. E.g., one can restrict the figures, or the manner in which "perimeter" is understood; or, in three dimensions, find the figure of greatest volume with fixed surface area. p. 213.

M

mean velocity If a particle travels a certain distance in a certain time interval, then the mean velocity with respect to that time interval is equal to the distance traveled divided by the length of the time interval. p. 180.

median In geometry, a median of a triangle is a line segment from a vertex to the midpoint of the opposite side. In statistics, the median of a set of numerical data is the middle value of the data. There are equally many data points above the median as there are below it. p. 98.

N

natural number A positive whole number. p. 5.

O

osculating circle The osculating ("kissing") circle at a point on a curve is the circle that fits the curve most closely at that point. (Recall that three noncollinear points—three distinct points not in a straight line—determine a circle.) The osculating circle at P can be approximated as close as desired by circles that pass through triples of points on the curve close to P. p. 63.

P

partial quotient The sequence of **natural numbers** that defines a simple continued fraction. For the simple continued fraction

$$\cfrac{1}{6 + \cfrac{1}{5}}$$

the numbers 6 and 5 are partial quotients. p. 27.

prime factorization The representation of a **natural number** as a product of **prime numbers**, for example, $60 = 2^2 \cdot 3 \cdot 5$. According to the *unique factorization theorem*, every **natural number** greater than 1 can be expressed as a product of **prime** numbers in exactly one way (apart from the order of the factors). p. 16.

prime number A **natural number** greater than 2 that is divisible only by itself and 1. p. 16.

principal curvature In general, a point P on a surface S has two principal curvatures, the maximum and the minimum curvatures at P of curves on S containing P. The product of the principal curvatures is the **Gaussian curvature** of S at P. p. 66.

proportion An equality of two **ratios**. The proportion asserting the equality of **ratios** $a : b$ and $c : d$ is written $a : b :: c : d$. p. 22.

R

radical **1.** A square **root**, cube root, nth root, etc. **2.** The radical sign $\sqrt{}$. p. 129.

ratio The relative size of two magnitudes. The ratio $a : b$ of two numerical quantities is defined by the fraction a/b (assuming that b is not equal to zero). p. 21.

rational number A number expressible as the quotient of two **integers**, for example, $7/5$. p. 5.

reciprocal The reciprocal of a number n is equal $1/n$. E.g., the reciprocal of 6 is $1/6$. p. 6.

relatively prime Two **natural numbers** are said to be relatively prime if their greatest common divisor is equal to 1. p. 27.

root **1.** A number that solves an equation. E.g., $x = 3$ is a root of $x^2 + 10x - 39 = 0$. **2. Square root, cube root, nth root, etc.** A number whose square, cube, nth power, etc., is equal to a given number. The square roots of 4 are 2 and -2 (that is, ± 2). The *positive* nth root of a number a (if it exists) is denoted $\sqrt[n]{a}$. p. 116.

S

sexagesimal system The representation of numbers as a sum of powers of 60. E.g., in the sexagesimal system, $11, 0, 21; 12, 45$ represents the number

$$11 \cdot 60^2 + 0 \cdot 60 + 21 + \frac{12}{60} + \frac{45}{60^2}$$

Using hours, minutes, and seconds, time is represented in the sexagesimal system. p. 15.

sharp An inequality involving $\leq$ or $\geq$ is said to be sharp if the inequality sign can be replaced by equality for some choice of the variables. E.g., the inequality $x^2 + y^2 \geq 2xy$ is sharp because equality holds if $x = y$. p. 223.

sphere In mathematics, a sphere is a certain *surface*, exclusive of its interior. A sphere, together with its interior, is called a *ball*. p. 67.

T

taxonomic key A hierarchical tree of questions that achieves the classification of an item belonging to a given **taxonomy**. p. 158.

taxonomy A hierarchical system of classification.

theorema egregium The theorem of Gauss that asserts that the **Gaussian curvature** of a surface is an intrinsic property. p. 66.

U

unit fraction A fraction with numerator equal to 1, for example, 1/5. p. 6.

V

variational principle One of a class of physical laws that assert, in some limited context, that of all conceivable outcomes, the one that actually happens maximizes or minimizes a certain quantity; e.g., the *principle of least action*. p. 212.

W

waveform A graphical representation of a (periodic or nonperiodic) wave. An acoustical waveform is a graph of pressure against time. p. 39.

Notes

Chapter 1 *Ancient Fractions*

1. Leonardo of Pisa is also known as Fibonacci. He is best known for the Fibonacci sequence: $1, 1, 2, 3, 5, 8, 13, 21, \ldots$. He used this sequence to describe the population growth of a colony of rabbits. He introduced the Hindu–Arabic place-valued decimal system of numerals to Europe.

2. A more sophisticated theory of fair division can show how to divide a (non-fungible) pizza among people who have different notions of the desirability of anchovies versus pepperoni.

3. Van der Waerden (1975) gives a more complete account of the ancient Egyptian arithmetic techniques.

4. Benson (1999, p. 101).

5. Van der Waerden (1975) discusses several such patterns.

Chapter 2 *Greek Gifts*

1. Translated by G. R. G. Mure.

2. For example, see Hardy and Wright (1979).

3. Translated by T. L. Heath.

4. In modern terms, we would define ratios as *equivalence classes*.

5. Fowler (1987) supports this thesis. Van der Waerden (1975, pp. 176–7) also supports an important role for anthyphairesis in ancient Greek mathematics.

6. Dedekind's unique further contribution was the concept of completeness and the foundations of the arithmetic of the real numbers.

7. Or, perhaps, the *Euclid team*. Nothing is known of Euclid's *curriculum vitae* — not even the dates of his birth and death. His role in the authorship of the *Elements* is unclear.

8. It can be argued that the standard term *division algorithm* is inappropriate because this procedure involves no division — only repeated subtraction.

9. Despite the fact that no division is involved, we use the term *partial quotient* because that is the standard term in the theory of continued fractions.

10. Euclid's *Elements*, **Book VII, Proposition 1.** *When two unequal numbers are set out, and the less is continually subtracted in turn from the greater, if the number which is left never measures the one before it until a unit is left, then the original numbers are relatively prime.*

Proposition 2. *To find the greatest common measure of two given numbers not relatively prime.*

11. Euclid's *Elements*, **Book X, Proposition 1.** *Two unequal magnitudes being set out, if from the greater there be subtracted a magnitude greater than its half, and from that which is left a magnitude greater than its half, and if this process be repeated continually, there will be left some magnitude which will be less than the lesser magnitude set out.*

Proposition 2. *If, when the less of two unequal magnitudes is continually subtracted in turn from the greater that which is left never measures the one before it, then the two magnitudes are incommensurable.*

12. See Fowler (1987) and Van der Waerden (1975).

13. However, the *completeness* of the real numbers was not understood by the

Greeks. Furthermore, the Greeks gave ratios an *order*, but they were much less clear concerning the *arithmetic* properties of ratios.

14. It can be shown by truncating the continued fraction—using a sufficiently large but finite number of the partial quotients—we obtain arbitrarily close approximations of $\sqrt{2}$.

Chapter 3 *The Music of the Ratios*

1. Aristotle, *Metaphysics*.

2. A diatonic scale consists of seven distinct tones: *do re mi fa sol la ti*.

3. For an ideal spring, the restoring force is proportional to the displacement. The vibrating spring is an example of *simple harmonic motion*.

4. Decibels are related to pressure amplitude by the formula $d = 10 \log p$, where p and d denote pressure and decibels, respectively. The unit of pressure is defined such that one unit represents a barely audible sound.

5. The details of this approximation are addressed by the theory of Fourier series.

6. The sum of waveforms is obtained by adding *vertical* dimensions.

7. Ohm is also known for Ohm's law of electrical resistance.

8. "The advancement and perfection of mathematics are intimately connected with the prosperity of the State" Napoleon Bonaparte, 1769–1821.

9. Helmholtz was a scientist of remarkable breadth. He made important contributions in physiology, optics, acoustics, and electrodynamics. Helmholtz showed the power of physical science in the study of sight and hearing. See Helmholtz (1954).

10. Musically unsophisticated subjects were sought for this experiment. It was thought that musically trained subjects would give undue weight to intervals that they were trained to hear.

11. A semitone is the interval between two adjacent notes of the chromatic scale, for example, C–C♯.

12. See also Schechter (1980).

13. Here is the *exact* calculation of the BAFS of the ratio $\log 2 : \log 3/2$.

$$\log 2 - 1 \cdot \log (3/2) \quad = \log 2/(3/2) = \log (2^2/3)$$
$$\log (3/2) - 1 \cdot \log (2^2/3) \quad = \log (3/2)/(2^2/3) = \log (3^2/2^3)$$
$$\log (2^2/3) - 2 \cdot \log (3^2/2^3) \quad = \log (2^2/3)/(3^4/2^6) = \log (2^8/3^5)$$
$$\log (3^2/2^3) - 2 \cdot \log (2^8/3^5) \quad = \log (3^2/2^3)/(2^{16}/3^{10}) = \log (3^{12}/2^{19})$$
$$\log (2^8/3^5) - 3 \cdot \log (3^{12}/2^{19}) = \log (2^8/3^5)/(3^{36}/2^{57}) = \log (2^{65}/3^{41})$$
$$\log (3^{12}/2^{19}) - 1 \cdot \log (2^{65}/3^{41}) = \log (3^{12}/2^{19})/(2^{65}/3^{41}) = \log (3^{53}/2^{84})$$

14. See Hardy and Wright (1979).

15. Piano tuners "stretch" the tuning of the upper and lower octaves of the piano. This means that they tune the upper octaves slightly sharp and the lower octaves slightly flat. It is generally agreed that this adjustment improves the sound of the piano. It is thought that this happens because real piano strings, due to their stiffness, do not vibrate exactly like an ideal vibrating string.

16. Silver (1971).

17. Jeans (1937) refers to use of continued fractions for equal-tempered tuning and notes that a scale of 53 equal intervals has an additional incidental advantage: The 53-tone scale gives an excellent approximation to the interval associated with the ratio 5 : 4, the major third according to *just* tuning.

Chapter 4 *Tubeland*

1. *Sphereland*, Burger (1965).

2. By a consortium led by Andrew Lange of the California Institute of Technology and Paolo Bernardis of Universitá de Roma, "La Sapienza."

3. By a University of California team led by Paul L. Richards.

4. In ordinary usage, one might say the surface in Figure 4.3(a) is convex when viewed from above and concave when viewed from below. This distinction is not important in our current discussion.

5. This is a real theorem of Euclidean geometry, not just a fictional one.

6. In modern notation, if the sides of the top and bottom squares are a and b, then the volume is given by $h(a^2 + ab + b^2)/3$.

7. Let the consecutive sides of a quadrilateral be $a, b, c,$ and d. For the area, the Egyptians used the incorrect formula $(a + b)(c + d)/4$.

8. As we will see, one of the axioms, the *parallels axiom*, is much less self-evident than the others.

9. Euclid, himself, occasionally erred in this respect. In fact, the foundations of geometry as presented in *The Elements* have certain flaws. Using Euclid's axioms as he wrote them, certain proofs make tacit use of geometric intuition—despite Euclid's requirement to the contrary. These shortcomings—dealing, for example, with the concepts of *betweenness* and *continuity*—have been corrected, for example, in Hilbert (1902). By *Euclidean geometry* we mean the geometry of Euclid *with all the necessary additions and corrections*.

10. In fact, Euclid does give definitions of point, line, and plane. For example, Book I, Definition 1 reads, "A point is that which has no part." From the modern point of view, this is not a definition because it does not define *point* in terms of previously defined terms. In fact, since this is the very first definition, there are no previously defined terms.

11. This form of the parallels axiom is known as *Playfair's axiom* after the Scottish mathematician John Playfair (1748–1819), uncle of the architect William Playfair (see page 89). It is more transparent than the equivalent version of the parallels axiom stated in Euclid's *Elements*.

Chapter 5 *The Calculating Eye*

1. For look at the many forms graphs may take, see Harris (1999). For beautiful and ingenious examples of the visual display of information, see Tufte (1983), (1990), and (1997).

2. Represented (without foundation) as a Chinese saying, this quotation was used in advertisements for *Royal Baking Powder*.

3. See Carter (1999).

4. In 1907, extensive development of the inferior parietal regions was already noted in the preserved brains of the mathematician C.F. Gauss and the physicist Siljeström (Witelson et al. (1999)).

5. Sir Jacob Epstein (1880–1959), U.S.-born English sculptor. Gertrude Stein (1874–1946), French-resident U.S. author.

6. See the essay by Henri Poincaré in Newman (1956, pp. 2041–2050), and Hadamard (1954).

7. Van der Waerden (1975, p. 118).

8. Outline of proof: Observe that the triangle with vertexes $(0,0)$, $(0,1)$, and (x,y) is a right triangle. The vertical line through (x,y) divides this large right triangle into two smaller right triangles, each similar to the large right triangle. Similarity implies that certain ratios hold, which imply the desired result.

9. Uncle of the noted Scottish architect William Henry Playfair (1789-1857) and brother of the mathematician John Playfair (1748–1819).

10. By 1933, the *New York Times* showed graphs of *Weekly* and *Daily Averages of 50 Combined Stocks*.

11. For a point below the X-axis or to the left of the Y-axis, the corresponding coordinate (x or y) is negative.

12. Smith (1959, pp. 389–402). Both Descartes and Fermat discovered analytic geometry in 1629, although Descartes published *La Géométrie* in 1637 and Fermat's work was published posthumously in 1679. We say that the analytic geometry of Descartes and Fermat is concerned with the graphs of equations; nevertheless, their diagrams did not include explicit coordinate axes.

13. Recall the mathematical meaning of the word *circle*. A circle is a curve — the circumference only. The circle, together with its interior, is called a *disk*.

Chapter 6 *Algebra Rules*

1. This quote is from the movie *The Wizard of Oz*, not from the book by Frank Baum.

2. Military theorist Karl von Clausewitz (1780–1831) said, "War is merely the continuation of politics by other means."

3. Hewlett–Packard calculators use a system called *postfix* instead of the order of precedence.

4. Euclid's proposition is slightly more general. It provides for division of the rectangle $\mathcal{R}$ into an *indefinite* number of subrectangles. This extension is proved by repeated application of the proposition as stated.

5. This equation is called *quadratic* because of the presence of the term x^2. Equations involving x^3 and x^4 are called *cubic* and *biquadratic*, respectively.

6. The square of -19.5 is also 380.25. This gives us a second solution of the equation $n(n-1) = 380$, namely, $n = -19.5 + 0.5 = -19$. However, the original statement of the problem does not permit a negative value for n.

7. A computer spreadsheet enables tabulation, manipulation, and calculation with rectangular arrays of data. Spreadsheet calculations are automatically updated as the data change. Although spreadsheets are widely used for business applications, they should be considered a general purpose computational tool. *VisiCalc*, the first commercial computer spreadsheet, became available to the public in 1979.

Chapter 7 *The Root of the Problem*

1. Recall that a number is rational or irrational depending on whether or not it can be expressed as the quotient of two integers. For example, $2/3$ is rational, and Proposition 2.1 shows that $\sqrt{2}$ is irrational. See page 21.

2. Recall that a complex number is of the form $a + ib$ where $i = \sqrt{-1}$ and a and b are real numbers.

Chapter 8 *Symmetry Without Fear*

1. The Alhambra is the Moorish fortress and palace overlooking Grenada, Spain. It was built in the twelfth and thirteenth centuries.

2. For a beautiful and historic collection of plane ornaments from many epochs starting with ancient Egypt, see Jones (1986) (originally published in 1856). For a discussion of the mathematical theory of plane ornaments together with a fine collection of examples from many cultures, see Washburn and Crowe (1988). M.C. Escher is the premier twentieth-century artist in this genre. See, for example, Escher (1961).

3. Based on Jones (1986), Plate XLV–21, 22.

4. (a) Based on Jones (1986), Plate XLII*–4. (b) Based on Jones (1986), Plate XLII†–5. These sources are more ornate than Figures 8.2(a) and (b).

5. A recent novel by Petsinis (1997) gives a fictionalized account of Galois's short life. Galois made fundamental contributions to mathematics before his absurd death in a duel at the age of 21.

6. An element that is its own inverse is said to be *idempotent*.

7. See Schattschneider (1978) for details of this naming convention and a detailed discussion of the wallpaper groups.

8. For example, see Munz (1970).

9. Figure 8.1(a): pma2. Figure 8.1(b): p1a1. Figure 8.2(a): p4m. Figure 8.2(b): p6.

Chapter 9 *The Magic Mirror*

1. Hilbert and Ackermann (1950).

2. Gödel (1931).

3. The twentieth century brought on many other influences pulling in the same direction: Einstein's theory of relativity, the mathematical theory of chaos, and the Heisenberg uncertainty principle.

Chapter 10 *On the Shoulders of Giants*

1. Indiana House Bill No. 246 (1897) reads in part, "Be it enacted by the General Assembly of the State of Indiana: it has been found that a circular area is to the square on a line equal to the quadrant of the circumference, as the area of an equilateral rectangle is to the square of one side." It is unclear what value for π is actually proposed here. The wording is unclear, but it probably implies $\pi = 4$.

2. Normality requires randomness of the sequence of, not only the base-10 digits, but even the digits with respect to each base. Furthermore, each finite subsequence of digits must recur randomly.

3. Thanks to G.D. Chakerian for this observation.

4. See Heath (1953, p. 4, Assumption 2), *On the Sphere and Cylinder*. The inside curve, the waist W, must be convex, but it is possible to weaken the assumption that the outer curve, the belt B, is also convex.

5. Figure A.4 provides a demonstration of this ratio. On the given 30°–60° right triangle OAB find point C such that the angle CAO is 30°. The two single-ticked angles are both 30°, and all three double-ticked angles are 60°. It follows that triangle OAC is isosceles and triangle ABC is equilateral. Therefore, the lengths of the ticked sides OC, CB, and AB are of equal length. By the Pythagorean

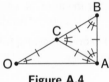

Figure A.4

theorem, if the length $\overline{AB}$ is equal to 1, then $\overline{OB} = 2$ and
$\overline{OA} = \sqrt{2^2 - 1^2} = \sqrt{3}$.

    6. An equation of a parabola is also mentioned on page 106.

    7. Quadratic equations are discussed on pages 130–135.

    8. Slope is discussed on page 99.

    9. *Secant* formerly denoted a straight line that meets a circle in two points, but now it also means a straight line that intersects an arbitrary curve in (at least) two points. *Secant* has a technical meaning in trigonometry — the reciprocal of the *cosine*.

    10. See Drake (1978) for an account of Galileo's scientific work including an account of the inclined plane experiment described below.

    11. See the discussion of constant speed on page 99. Speed is always a nonnegative quantity. Velocity is defined as speed together with the direction of motion. One-dimensional velocity is speed together with a plus or minus sign to denote the direction.

Chapter 11 *Six-Minute Calculus*

    1. Bishop George Berkeley (1685?–1753) was an outspoken critic. In *The Analyst: A Discourse addressed to an Infidel Mathematician*, he called Newton's fluxions "the ghosts of departed quantities."

    2. Euler was surely the most prolific mathematician of all time. Publication of his collected works, almost all in Latin, is an ongoing project that currently consists of 72 volumes.

    3. The arrow $\to$ in the formula $\lim_{x \to a} g(x) = L$ means "tends." The arrow $\mapsto$, used on page 187, means "maps to."

    4. Note that we are using mean velocity in the same sense as in the preceding chapter. (See Table 10.)

    5. Differentials can be given a modern rigorous meaning, but we will not do so in this book.

    6. For curvilinear motion of a particle, velocity has both magnitude and direction.

    7. The variable t in the expression $\int_0^T f(t)\, dt$ is called the *variable of integration*. This is an example of a *dummy* variable, so named because this formula does not depend on t. In place of t, we could have used u or any other variable.

    8. This assertion is a consequence of the *mean value theorem*:

Theorem (Mean value theorem). *Suppose that the function $G(t)$ is continuous on the interval $a \leq t \leq b$ and has a derivative $g(t)$ in the interior of that interval. Then there exists a number c $(a < c < b)$ such that $g(c)$ is equal to*

$$\frac{G(b) - G(a)}{b - a}$$

This fraction can be interpreted as a difference quotient or as the mean velocity over the interval $a \leq t \leq b$.

    9. The *second* fundamental theorem of calculus is the following:

Theorem (Second fundamental theorem of calculus). *Let $g(t)$ be continuous on the interval $a \leq t \leq b$. Define $G(t)$ by the formula*

$$\int_a^t g(s)\, ds$$

Then $G(t)$ is differentiable and $G'(t) = g(t)$ for all t in the interval $a < t < b$.

Chapter 12 *Roller-Coaster Science*

1. See page 196.

2. For help in computing derivatives, see Table 11.3 (table of derivatives) and Table 11.4 (rules for derivatives).

3. See Honsberger (1973).

4. This is an example of a standard topic of elementary algebra—*completing the square*.

5. Inequalities are the mainstay of mathematical analysis. The classic work on this subject is Hardy, Littlewood, and Pólya (1934).

6. The following is the general case of Cauchy's inequality:

Theorem (Cauchy's inequality—general case). *Let n be a natural number. For any real numbers $x_1, x_2, \ldots, x_n$ and $y_1, y_2, \ldots, y_n$, we have*

$$\left(\sum_{i=1}^{n} x_i y_i \right)^2 \leq \sum_{i=1}^{n} x_i^2 \, \sum_{i=1}^{n} y_i^2$$

Equality holds only if the numbers $x_1, x_2, \ldots, x_n$ and $y_1, y_2, \ldots, y_n$ are proportional. In other words, equality holds only if $x_i y_j = x_j y_i$ for all natural numbers i and j satisfying $1 \leq i, j \leq n$.

7. See Figure 5.3(b) on page 86.

8. See Twardokens (1990).

9. The definition of integral (12.11) entails a technical difficulty. Since y is zero when x is zero, the fraction

$$\frac{\sqrt{1 + y'^2}}{\sqrt{y}}$$

is undefined at $x = 0$ because division by 0 is undefined. In fact, this fraction becomes unbounded near $x = 0$. The integral (12.11) is an example of what is called an *improper* integral. For positive numbers ϵ, however small, the integral

$$\int_{\epsilon}^{x_0} \frac{\sqrt{1 + y'^2}}{\sqrt{y}} \, dx$$

is an ordinary integral. The limit, finite or infinite, of this integral as ϵ tends to zero through positive values is technically what is meant by the integral (12.11). Of course, an infinite value is not relevant to the brachistochrone problem because we already know that there are paths that produce finite descent times.

10. Calculating the particulars of this cycloid is beyond the scope of this book. Approximate numerical methods were used.

References

Abbott, E. A. (1884). *Flatland: A romance of many dimensions*. London: Seeley.

Benson, D. C. (1969). A elementary solution of the brachistochrone problem. *American Mathematical Monthly, 76*(8), 890–894.

Benson, D. C. (1999). *The moment of proof: Mathematical epiphanies*. New York: Oxford University Press.

Burger, D. (1965). *Sphereland*. New York: Crowell. (Translated from the Dutch by Cornelie J. Rheinboldt. *Sphereland* is a sequel to *Flatland*, Abbott (1884).)

Carter, R. (1999). *Mapping the mind*. Berkeley: University of California Press.

Drake, S. (1978). *Galileo at work*. Chicago: University of Chicago Press.

Escher, M. C. (1961). *The graphic work of M. C. Escher*. New York: Meredith Press.

Fowler, D. H. (1987). *The mathematics of Plato's Academy: A new reconstruction*. Oxford: Oxford University Press.

Galilei, V. (1985). *Fronimo* (vol. 39). Neuhausen-Stuttgart: American Institute of Musicology. (Translated and edited by Carol MacClintock. First published in 1584. This work discusses the lute's temperament and the issue of alternative frets (tastini) at pp. 155–166.)

Gödel, K. (1931). Über formal unentscheidbare Sätze der Principia Mathematica und verwandter Systeme I (On formally undecidable propositions of Principia Mathematica and related systems I). *Monatshefte für Mathematik und Physik, 38*, 173–198.

Grant, E. (ed.). (1974). *A source book in medieval science*. Cambridge, MA: Harvard University Press.

Hadamard, J. (1954). *An essay on the psychology of invention in the mathematical field*. New York: Dover. (Reprint. First published by Princeton University Press in 1945.)

Hardy, G., Littlewood, J., and Pólya, G. (1934). *Inequalities*. Cambridge: Cambridge University Press.

Hardy, G., and Wright, E. (1979). *An introduction to the theory of numbers* (5th ed.). Oxford: Oxford University Press.

Harris, R. L. (1999). *Information graphics: A comprehensive illustrated reference.* New York: Oxford University Press.

Heath, T. L. (ed.). (1953). *The works of Archimedes.* New York: Dover. (Reprint. Originally published by Cambridge University Press in 1897.)

Helmholtz, H. L. F. (1954). *On the sensations of tone as a physiological basis for the theory of music.* New York: Dover. (Translation by Alexander J. Ellis of the fourth (and last) German edition of 1877.)

Hilbert, D. (1902). *The foundations of geometry.* Chicago: The Open Court Publishing Company. (Translated by E. J. Townsend.)

Hilbert, D., and Ackermann, W. (1950). *Principles of mathematical logic.* New York: Chelsea. (Translated from the German by Lewis M. Hammond, George G. Leckie, and F. Steinhardt. First published in 1928 with the title *Grundzüge der theoretischen Logik.*)

Honsberger, R. (1973). *Mathematical gems.* Washington, DC: The Mathematical Association of America.

Jeans, S. J. (1937). *Science and music.* New York: MacMillan.

Jones, O. (1986). *The grammar of ornament.* London: Studio Editions. (Reprint. Originally published by Messers Day and Son, London, in 1856.)

Kőnig, D. (1950). *Theorie der endlichen und unendlichen Graphen.* New York: Chelsea. (Reprint. Originally published in 1936.)

Kreith, K., and Chakerian, D. (1999). *Iterative algebra and dynamic modeling: A curriculum for the third millenium.* New York: Springer-Verlag.

Lambert, J. H. (1779). *Pyrometrie.* Berlin: Hauder and Spener.

Munz, P. A. (1970). *A California flora.* Berkeley: University of California Press.

Newman, J. R. (ed.). (1956). *The world of mathematics.* New York: Simon and Schuster.

Nightingale, F. (1858). *Notes on matters affecting the health, efficiency, and hospital administration of the British Army, founded chiefly on the experience of the late war.* London: Harrison and Sons.

Parshall, K. H. (1995). The art of algebra from al-Khwarizmi to Viète: A study in the natural selection of ideas. *World Wide Web.* (http://viva.lib.virginia.edu/science/parshall/algebra.html)

Petsinis, T. (1997). *The French mathematician.* New York: Berkley.

Playfair, W. (1801). *Commercial and political atlas.* London: Wallis.

Plomp, R., and Levelt, W. (1965). Tonal consonance and critical bandwidth. *Journal of the Acoustical Society of America, 38(2),* 548–560.

Reid, D. A. (1999). Symmetry in the plane. *World Wide Web.* (http://plato.acadiau.ca/courses/educ/reid/Geometry/Symmetry/symmetry.html)

Schattschneider, D. (1978). The plane symmetry groups: Their recognition and notation. *American Mathematical Monthly, 85,* 439–450.

Schechter, M. (1980). Tempered scales and continued fractions. *American Mathematical Monthly, 87*(1), 40–42.

Schulter, M. (1998). Pythagorean tuning and medieval polyphony. *World Wide Web.* (http://www.medieval.org/emfaq/harmony/pyth.html)

Silver, A. L. (1971). Musimatics or the nun's fiddle. *American Mathematical Monthly, 78*(4), 351–357.

Smith, D. E. (1959). *A source book in mathematics* (vol. 2). New York: Dover.

Tufte, E. R. (1983). *The visual display of quantitative information.* Cheshire, CT: Graphics Press.

Tufte, E. R. (1990). *Envisioning information.* Cheshire, CT: Graphics Press.

Tufte, E. R. (1997). *Visual explanations : Images and quantities, evidence and narrative.* Cheshire, CT: Graphics Press.

Twardokens, G. (1990). Brachistochrone (that is, shortest time) in skiing descents. *Proceedings of the Eighth International Symposium of the Society of Biomechanics in Sports,* 205–209.

Van der Waerden, B. L. (1975). *Science awakening* (vol. 1). Groningen: P. Noordhoff. (Originally published by P. Noordhoff in 1961 as *Ontwakende wetenschap.*)

Walter, M., and O'Brien, T. (1986). Memories of George Pólya. *Mathematics Teaching, 116.*

Washburn, D. K., and Crowe, D. W. (1988). *Symmetries of culture.* Seattle: University of Washington Press.

Whittaker, E. T., and Watson, G. N. (1927). *A course of modern analysis* (4th ed.). Cambridge: Cambridge University Press. (The first edition was published in 1902.)

Witelson, S. F., Kigar, D. L., and Harvey, T. (1999). The exceptional brain of Albert Einstein. *The Lancet, 353*(9170), 2149–2153.

Index